29, 95

Y0-ABP-945

BIOLOGICAL DIFFERENCES
AND
SOCIAL EQUALITY

BIOLOGICAL DIFFERENCES
AND
SOCIAL EQUALITY

Implications for Social Policy

Edited by
Masako N. Darrough
and
Robert H. Blank

Foreword by Gardner Lindzey

GREENWOOD PRESS
WESTPORT, CONNECTICUT • LONDON, ENGLAND

Library of Congress Cataloging in Publication Data
Main entry under title:

Biological differences and social equality.

Papers based on a Summer Institute on Biological
Diversity and Social Equality held at the Center for
Advanced Study in the Behavioral Sciences at
Stanford during July and August 1978.
  Bibliography: p.
  Includes index.
  Contents: The reality and significance of human
races / Richard Goldsby — Search for equity on the
planet difference / Nancy Hauserman — Evolution,
ethics, and equity / Stephen L. Zegura, Stuart C.
Gilman, Robert L. Simon — [etc.]
  1. Equality—Congresses. 2. Social justice—
Congresses. 3. Nature and nurture—Congresses.
4. Sociobiology—Congresses. I. Darrough, Masako N.
II. Blank, Robert H. III. Summer Institute on
Biological Diversity and Social Equality (1978 :
Center for Advanced Study in the Behavioral
Sciences)
HM146.B56 1983   304.5   82-11914
ISBN 0-313-23022-6 (lib. bdg.)

Library of Congress Catalog Card Number: 82-11914
ISBN: 0-313-23022-6

First published in 1983

Greenwood Press
A division of Congressional Information Service, Inc.
88 Post Road West, Westport, Connecticut 06881

Printed in the United States of America

10 9 8 7 6 5 4 3 2 1

Copyright Acknowledgments
  The editors and publishers are grateful for permission to reprint from
the following works.
  Stephen C. Pepper, *World Hypotheses* (Berkeley: University of California
Press, 1966). Copyright © 1966 by The Regents of the University of California.
  Robert Simon and Stephen Zegura, "Sociobiology and Morality," *Social
Research* 46, no. 4 (1979). Copyright © 1979 by the New School for Social
Research.

Contents

Foreword

The intersection of biological differences and social equality presents one of the fundamental dilemmas of modern democratic society. Certainly the scientific tangle of environmental and biological determinants of behavior and the moral concern for maximizing for each individual the potential for accomplishment and personal fulfillment pose a public policy question of enormous complexity. While this collection of essays will surely not resolve the major scientific and policy issues involved, it represents as varied a set of thoughtful perspectives on the problem area as one could hope to find. The contributors to the volume cover the full array of behavioral sciences and, indeed, overflow into biological sciences and legal scholarship. Perhaps the only attribute the contributors have in common, other than an interest in biological variation and social equity, is youth.

The volume is an outcome, not originally planned, of one of a series of summer institutes arranged by the Center for Advanced Study in the Behavioral Sciences and largely funded by the Andrew W. Mellon Foundation. These institutes are intended to provide an opportunity for interdisciplinary faculty development and are aimed particularly at young and minority scholars and those affiliated with four-year colleges and regional universities. In selecting participants, the principal focus is upon demonstrated scholarship of high quality or very promising indicators of future research contributions. The institute itself involved a lively mixture of didactic presentations by the directors, consultants, and participants, as well as informal exchanges on a wide variety of topics. The interdisciplinary character of the institute is clearly underlined by the disciplinary backgrounds of the two directors—philosophy and psychobiology.

The center's primary function is the conduct of a residential postdoctoral fellowship program for distinguished scientists and scholars from this country and abroad. The summer institutes have enriched our general program and at the same time the institutes have benefitted from the presence of our regular fellows. Given our commitment to interdisciplinary and collaborative research, we are particularly pleased to have been involved with the activities that led to the diverse collection of papers that comprise this interesting volume.

Gardner Lindzey
Director, Center for Advanced Study
in the Behavioral Sciences

Preface

In *The Republic*, Plato tells us that by nature individuals have different abilities and worth, or, at the very least, are naturally suited to fulfill different roles and carry out different functions. The just society, he concludes, is the one in which individuals actually fill the roles for which they are best suited. The result is a kind of meritocracy with a vengeance. Given different natural endowments, and subjected to similar early environments, individuals will wind up at that level on the hierarchy for which their talents fit them. The rulers will be those best fitted to rule, the laborers those best fitted to labor, and so on.

Plato's *Republic* is likely to evoke ambivalent feelings in most readers. His belief in significant innate differences among humans, and his willingness to assign such great significance to them, will trouble those with even moderately egalitarian sympathies. On the other hand, his emphasis on talent and ability, rather than race, social class, or even sex, may tend to strike a responsive chord in many readers.

But regardless of our reaction to it, *The Republic* raises two of the great questions about equality. How great are the innate or biological differences among humans? What social significance should be assigned to innate or biologically grounded factual inequalities?

Or, in other words, how unequal are we, and what if anything should be done about any natural inequalities which might exist?

These questions clearly are alive today and arise in a number of areas of social controversy. Are differences in performances between individuals and between groups on some standardized tests the result of unequal environments or of biological inequalities? Is such a question even susceptible to analysis, and if so what are the implications for social policy?

Factual issues and normative ones tend to blur together in this area, however important it may be to try to keep them distinct. For example, the gathering of factual data of certain kinds may raise ethical issues of invasion of privacy. This may even involve stigmatization of individuals as where genetic screening is performed for race- or sex-linked disorders. Moreover, it is far from clear what social justice demands with respect to equality. Should we view society, as John Rawls has suggested, as if we were behind a "veil of ignorance" which prohibits us from knowing any of our personal characteristics? Is justice blind, then, to the outcome of the genetic lottery? Is the Rawlsian principle that inequalities are just only when they maximize the position of the disadvantaged the position we would choose from the Archimeadean point of Rawlsian impartiality?

These questions, factual and normative, can hardly be ignored. Consider questions of equality in education. Why do some children perform differently than others? What should society do about it? Surely, differences due in part to racial discrimination are unjust. But is equality of educational opportunity equivalent simply to nondiscrimination, as elusive as that ideal has proved to attain? Are differences in outcome acceptable if they arise simply from differences in ability? What if certain family backgrounds give some children an advantage over others, regardless of race? Should schools try to equalize outcomes, perhaps by paying special attention to the disadvantaged? What then of differences arising from differential natural ability? We seem to be facing once again problems similar to those which have occupied social theorists from Plato to Rawls.

It is questions such as these which were explored during July and August of 1978 by the participants in the Summer Institute on Biological Diversity and Social Equality held at the

Center for Advanced Study in the Behavioral Sciences at
Stanford. The essays in this collection arose out of the lectures,
seminars, and discussions which took place during the institute.
The participants included scholars from the following disciplines: anthropology, biology, economics, education, history,
law, medicine, philosophy, political science, psychology, and
sociology. A major goal of the institute was to provide a cross-
disciplinary investigation into significant social issues involving
human equality. We hope that publication of these essays will
show that goal to be a worthy one by illustrating how questions
of equality can be illuminated when looked at from a variety
of disciplinary perspectives.

   On behalf of those who participated in the institute, we wish
to thank the administration, staff, and fellows of the center
for providing an idyllic setting for inquiry and debate.

Baruch S. Blumberg
Institute for Cancer Research
Consultant, Summer Institute on Biological
Differences and Social Equality

Robert L. Simon
Hamilton College
Co-Director, Summer Institute on Biological
Differences and Social Equality

Kurt L. Schlesinger
University of Colorado
Co-Director, Summer Institute on Biological
Differences and Social Equality

I

THEORETICAL
DISCUSSION

1

The Reality and Significance of Human Races: A Biological Perspective

RICHARD A. GOLDSBY

The purpose of this essay is to provide a basic introduction to the biological reality, meaning, and significance of "race." In our attempt to illuminate the relationship between "biological differences" and "social equality," one of the essential tasks to be undertaken is to develop an understanding of the first of the two concepts which are to be explored in relation to one another. It will be shown here that there exists substantial genetic diversity both within and among races.

A race is a breeding population of individuals identifiable by the frequency with which a number of inherited traits appear in that population. A breeding population is one which for reasons of geography or culture mates largely within itself. Races can be recognized in all groups of living things, including humankind, that are widely distributed and have been systematically studied.[1] As examples of breeding populations that are recognized as human races, such groups as the Australian aborigine and the American black can be cited. It is important to recognize at the outset that racial characterizations are statistical descriptions of the way in which breeding populations of the same species differ from each other with respect to certain inherited traits. It should be stressed that it is not the uniformity with which a set of inherited characteristics appear in a population but rather the comparative frequency in one population of a species as opposed to other populations of the same species that serves to define and identify a population as a racial group. This chapter will survey some of the contributions, both factual and conceptual, various areas of biology have made to an appreciation of the reality and significance of race in human populations.

The biology of human racial populations is studied for the following three reasons: (1) the recognition and classification of human racial groups, (2) an understanding of the origin of human racial populations, and (3) the determination of the significance of racial differences for the conduct of human affairs. In this chapter consideration will be given to each of these three major areas of study.

The Classification of Human Races

One of the earliest attempts at the classification of human racial types was made by the Swedish biologist Carolus Linnaeus, who, in 1758, founded modern taxonomy with the publication of *Systema Naturae*. In formulating a system of racial classifications he incorporated a few genetic traits such as skin color, hair form, and hair color, and many nonbiological personal assessments. His system recognizes four racial groups within the species *Homo sapiens*.

1. *Homo sapiens americanus*, red, hot-tempered, erect, straight, black, thick hair, broad-nostriled, resolute, free, paints self with artful lines of red, is ruled by custom.
2. *Homo sapiens europaeus*, white, brawny, long, flaxen hair, blue-eyed, nimble, optimistic, of the keenest mind, innovative, is ruled by ritualistic tradition.
3. *Homo sapiens asiaticus*, pale yellow, meditative (gloomy), hardy, dark hair, dark-eyed, grave, proud, greedy, is ruled by opinions.
4. *Homo sapiens afer*, black, loose-limbed, kinky and dull black hair, flat nose, big lips, cunning, lazy, careless, sluggish, smears self with fat, is ruled by authority.

Modern systems for the classifications of human races differ sharply from the Linnaean one, have no place for characterization based on a group's religion or its real or imagined social psychology. For the definition and classification of human races only genetically determined traits are appropriate markers. The list of inherited traits used include such surface features as skin color, eye cast, hair form, and bone structure, to which are added blood type and other inherited details of body chemistry. The fact that internal indices such as blood type are of determinative importance in recognizing and differentiating human racial populations testifies to the reality of human racial differences and demonstrates that they can be assessed by means of objective laboratory examination.

Before presenting a classification of human racial groups it is useful to remind the reader that race is defined in terms of populations and the frequency with which certain inherited traits occur in particular populations as contrasted with other populations of the same species. A race is a breeding population identifiable by the frequency with which a number of inherited traits appear in that population. Consequently, it is not necessarily the uniformity with which a particular inherited characteristic appears in a group but rather its comparative frequency that serves to define a racial population. This means that not every member of a race necessarily will have all of the characteristics that collectively identify his or her racial population. As an illustration consider the following example. Occasionally (about once in every 20,000 births) in populations of African blacks an albino child completely lacking in pigment and, therefore, white is born. Even though it does not have the dark skin characteristics of the African black races, the child is, nevertheless, a member of an African black racial population.

There is much more variation within human racial populations than there is within races of domestic animals. Races of dogs (German shepherds, collies, dachshunds), cows (jerseys, holsteins, and guernseys), or chickens (white leghorns or Rhode Island reds) have been fashioned by the art and science of animal husbandry to fill practical or aesthetic needs. These arranged races are to a large degree "pure" (that is to say very uniform) because the breeder tends to prune variants from the population and to restrict mating opportunities to members of the breed. On the other hand, races such as those of sparrows, mice, iguanas, and humans that arise in nature, usually through the geographical isolation of various populations of the same species from each other, interbreed with members of different racial populations whenever the opportunity to do so occurs. Consequently, one finds that races in a natural state, as opposed to the rigidly genetically isolated races of domestic breeds, tend to be separated by intergrading zones rather than by sharp lines of demarcation. It is this dynamic genetic exchange between populations that leads to the variation always seen within racial populations that exist in a state of nature, outside the controls of artificial selection imposed by the animal or plant breeder. With these caveats in mind we can consider the classification of the human species into a number of racial groups recognizable by the frequencies with which inherited external and internal characteristics may be found within the populations.

The Mongoloid Races

Skin color in the Mongoloid races ranges from nearly white through an apparent yellow to brown. Generally speaking, the lighter skin variations are found to be more prevalent in the north, whereas the darker-skinned groups tend to be found in southerly latitudes bordering the tropics. The hair of Mongoloids tends to be black, of large diameter, round, and although hair may grow quite long on the head, heavy beards and generously distributed body hair are uncommon among the Mongoloid populations. The faces are characterized by high prominent cheekbones and almond-shaped eyes that are brown to black. A wide range of statures is found within these populations with some subgroups being quite tall whereas others average only a little over five feet. To these general characteristics a special anatomical feature that is distinctive to these groups should be added. The inside surface of the incisors of many members of these groups are concave or "shoveled." Five major subgroups are represented within the Mongoloid aggregate:

1. North Chinese: Northern and Central China and Manchuria
2. Classic Mongoloid: parts of Siberia, Mongolia, Korea, and Japan

3. Southeast Asian: South China, Thailand, Vietnam, Burma, and parts of Indonesia
4. Tibetan: Tibet
5. The American Indian group: the indigenous population of North, South, and Central America

The Caucasoid Racial Groups

Among the Caucasoid groups one finds a wide range of skin color varying from the pale translucent alabaster white of the Scandinavian through the Mediterranean tan of the Greek to the brown of the Arab. Similarly, a wide range of eye color is found among Caucasoid groups ranging from the blue so common in Sweden to the dark brown characteristic of the Greek. The hair of Caucasoids, which can be found in a spectrum of colors—yellow, red, brown, black—is usually straight or wavy. Although there is great variation in the structure of the nose, ranging from high and narrow to broad and snub, the lips are usually thin. Many of the males of this group can grow heavy beards, and there is a relatively large amount of body hair in both sexes. Among the Caucasoid racial groups one finds considerable variation in body builds from the medium short stature of Mediterranean populations to the tall rangy build characteristic of Scandinavian groups. The four major subcategories of these groups recognized are:

1. Northwest European: Scandinavia, Northern Germany, the Netherlands, Great Britain, and Ireland
2. Northeast European: Russia and Poland
3. Alpine: France, Southern Germany, Switzerland, Northern Italy, Yugoslavia, and the Balkans
4. Mediterranean: peoples from both sides of the Mediterranean from Tangiers to the Dardenelles including Arabia, Turkey and Iran

The Negroid Group

The Negroid group is characterized by color and body build. Skin color ranges from brown to black as does that of the hair and eyes. The hair on the head is often tightly curled and the body hair is sparsely distributed. Full lips, broad noses, small close-set ears and relatively rounded heads are common in this group. Within this group one finds a greater spread of height than encountered in either the Mongoloid or Caucasoid racial aggregates. Included within the Negroid aggregate are the towering Watusi, some of whom are as tall as seven feet, and the Pygmies, small people of the forest whose adult height is under five feet. This aggregate may be divided into the following subpopulations:

1. Sub-Saharan African: West African black, West African, and much of the Congo.
2. Bantu: Mozambique, Angola, parts of the Republic of South Africa and East Africa.
3. East African black: Kenya, Tanzania, parts of the Sudan, and Ethiopia.
4. The Forest Pygmy: the rain forests of equatorial Africa.
5. The North American black: the population resulting from an African Black-European hybridization, the majority of whose genes (approximately 80 percent) come from African black populations. In this population there is no evidence of a substantial contribution from the Indian populations indigenous to North America.
6. The South American black: this group, in common with the North American black, has a genetic background that is predominately African black, but in addition to the European admixture, there is a substantial contribution from South American Indian populations as well.

Races of the Indian Subcontinent

Skin color in this group ranges from light through medium brown to black. The hair is straight or wavy and black. Eyes are brown to black, lips thin to medium full, and cheekbones are distinct. Although some populations within this racial aggregate are tall, particularly those of Northern India, body builds are usually of medium height. Within this major aggregate two populations (which could be further subdivided) can be recognized:

1. The people of Pakistan and much of India
2. The people of Southern India and the island of Sri Lanka

The Caploid Group

This group, which contains the bushmen of Africa, is a disappearing race of man among whom fewer than 100,000 individuals can be counted. Their stature ranges from five feet to just a little less than five and a half feet, and they have brown-yellowish skin and relatively flat faces. The hair is often tightly coiled into spiral tufts, and it is mostly confined to the head and beard and is sparsely distributed over the balance of the body.

The Australoids

The Australoid group is found in Australia and parts of New Guinea. Its members range in color from nearly black to medium and light brown. The hair can vary from tightly coiled to straight, is widely distributed over the body, and, although usually black, individuals with hair of lighter hue are not uncommon. This race includes a high frequency of individuals with deep-set eyes, a large prominent nose, and full lips. Australoids vary in stature from medium height of around five and a half feet down to about five feet.

Racial Differences—Internal Indicators

The foregoing classification of human populations into different racial groups was based on statistical differences in color, hair form, facial features, and body type. A moment's reflection will suggest that these visually observable differences in color, body build, and facial cast must have a deeper constitutional basis. Such obvious differences in observable structure must be more than skin deep and would be expected to be manifested in the internal structure and chemistry of the individual. One could imagine an experiment to demonstrate that a number of internal as well as external constitutional differences exist which allow one to recognize racial groups as certainly do differences in outward appearance. Suppose three hundred people—one hundred of them Chinese, one hundred black Africans, and one hundred Caucasians from northern Europe—split up into their various groups and each race entered a room of its own. Could we, sight unseen, determine which room held the Chinese, the blacks, and the whites? The answer is that we could, definitely, provided that each individual in these various groups were willing to help us by supplying the following: (1) blood samples from each individual; (2) information on how many individuals in each group could detect certain flavors, in particular that characteristic of the compound PTC (phenylthiocarbamide); (3) a sample of ear wax from each individual. The data which follow in the sections below will demonstrate that the bit of detective work we set for ourselves is easily and definitively accomplished. Our little experiment is a conclusive demonstration that race, unlike beauty, is more than skin deep.

Blood Type

It is well known that there are different blood types, the most familiar being types A, B, AB, and O. In addition these familiar blood types, the presence or absence of Rh factor also provides another important index for blood classification. Since blood types are inherited traits which vary between racial populations they provide excellent indices for the differentiation of racial populations. An examination of Table 1, which lists the blood group frequencies found in a number of racial populations, will demonstrate that the frequency of individuals possessing one or another of the blood types displayed varies from one racial population to another. If we examine Table 1, concentrating on the types A_2, Rh negative, and Duffy, we find that Caucasians

Table 1   **Frequencies of Selected Blood Groups in Different Populations (In Percentages)**

Population	A_1	A_2	B	O	Rh Negative	Duffy Factor
Caucasians	5-40	1-37	4-18	45-75	25-46	37-82
American Blacks	8-30	1-8	10-20	52-70	4-29	0-6
Asian Mongoloids	0-45	0-5	16-25	39-68	0-5	90-100
American Indians	0-20	about 0	0-4	68-100	about 0	22-99

SOURCE: see note 2.

have more Rh negatives than any other major racial group and they also have more A_2. American blacks have a lower frequency of Rh negatives than Caucasians but a significantly higher one than those found in Mongoloid populations. Note also that American blacks are distinctive in their low frequency of the Duffy factor. If one considers the proportions of Rh negatives, type A_2, and Duffy positives, the Asian Mongoloid aggregate is readily identified. This group, like the American Indian, has few Rh negative individuals. Among Asian Mongoloids there is a high frequency of Duffy positives and fewer A_2 individuals than among Caucasians and American blacks. The American Indian is readily differentiated on the basis of the frequencies of type O individuals and the much lower frequency of type B.

Tasters and Nontasters

Many people are able to taste the bitter flavor of PTC, a compound that along with some chemically related substances has antithyroid activity. If bits of filter paper are impregnated with a dilute solution of PTC and given to subjects to determine if they can taste its slightly bitter flavor, the majority of individuals in any racial population can detect its bitter taste. However, the size of the majority tends to vary among racial groups as shown in Table 2.

Table 2   Frequencies of PTC Tasters in Different Populations
(In Percentages)

Population	PTC Tasters
European Caucasoids	60-80
American Blacks	90-97
Asian Mongoloids	83-100
Australian Aborigines	50-70

SOURCE: see note 3.

One finds the lowest percentage of tasters among Australian aborigines and the highest percentage among racial populations of the Mongoloid aggregate and of the black aggregate.

Ear Wax

The waxy secretion of the ear comes in two forms, one crumbly and dry and the other moist and adhesive. Whether or not an individual has dry, crumbly or sticky, adhesive ear wax is determined genetically. As shown in Table 3 the variations in the consistency of ear wax seen when one compares different racial populations demonstrates that it is possible to differentiate population groups based on this parameter. Clearly, populations

Table 3   Frequency of Dry Ear Wax Gene in Different Populations (In Percentages)

Population	Dry Ear Wax Gene	Standard Error
Northern Chinese	97.9	0.7
Southern Chinese	86.0	1.0
Japanese	91.5	0.1
Caucasoids of Germany	17.6	2.2
Caucasoids of U. S. A.	15.8	4.2
American Blacks	6.9	3.2

SOURCE: see note 4.

of the Mongoloid aggregate possess a very high percentage of genes stipulating the dry and crumbly form of ear wax. In contrast, populations stemming from the Caucasoid or black aggregates have low frequencies of individuals with dry ear wax and high frequencies of individuals with sticky and adhesive ear wax.

Genetic Differences between Races As Compared to Genetic Differences within Races

It is clear that average differences in genetic makeup can be readily demonstrated when comparing one racial group with another. Some perspective on the significance of these inter-group differences can be gained by exploring the average differences seen within racial groups. Suppose we have two Caucasoid individuals, W and w, picked at random and two blacks, B and b, selected in a similarly random fashion. It is appropriate to ask what is the magnitude of the genetic difference between W and w or between B and b as compared to the average difference between Caucasoid and black groups. The differences in the chemical structure of the proteins produced by two different individuals, or the average differences in the structure of the proteins produced by two different racial populations, provides a direct answer to the question posed. This is because every gene of an organism specifies the detailed structure types of proteins that play their various roles, as diverse as digestion and defense, in maintaining the integrity of the body. Consequently, differences in proteins taken from two different sources can be used as a measure of the genetic differences between these sources. Such analyses have shown (Cavalli-Sforza, 1974) that an average difference of about 0.02 percent will be found between members of the same racial population. The average difference between two racial groups such as American blacks and whites or whites and Chinese will be on the order of slightly more than 0.02 percent, just a little more than the variation between members of the same racial group. These considerations show that there is considerable variation between human beings whether they belong to the same racial group or not, and they also constitute a demonstration that racial differences, though real, add very little to the variation that already exists between human beings.[5]

The Problem of Racial Origins

The fossil record which has been so helpful in tracing man's descent from distant ancestors cannot be used to gain information on the origin of most contemporary human races. Consider for a moment the difficulties encountered in making a rigorous differentiation between fossils of *Homo erectus* and *Homo sapiens*, two different species of man separated in time. One of the most important characteristics on which to base this distinction is brain size, a generalization that remains true even though complicated by the fact that there is some overlap in size between the larger cranial fossils left by *Homo erectus* and some of the smaller skulls found in populations of *Homo sapiens*. Thus even where marked differences in populations exist, it can be difficult to separate the fossils of one group, unequivocally, from those of the other. Consider also the fact that the most magnificent fossils are at best only heaps of once interconnected bones. There is no hair, no skin, no flesh, and no blood. Consequently, no amount of careful study of even the most perfect human fossils is going to allow one to determine the racial identity of the populations which left the fossils. Add to these complications the consideration that in order to establish a pattern of evolution for a particular race, we would have to be able to see a gradation in time of racial characteristics that would culminate in one of today's existing racial types. To be useful in tracing racial origins the fossil record would be required to show not only clear racial differences, but also the small and subtle divergence in time of group differences during the evolutionary ascent of a race. The fossil record cannot do this.

Nevertheless, it is possible to discuss the ways in which breeding populations characterized by certain gene frequencies that are different from those found in other breeding populations of the same species are formed and maintained. Such a discussion requires an understanding of what is necessary to establish and maintain differential gene frequencies in different populations.

The forces of variation, selection, and isolation are important in race formation. Without the occurrence of natural genetic variations in populations the raw material for the selection of new lines does not exist, and evolution is precluded.[6] Some degree of genetic isolation is necessary in order to prevent populations differing in the frequency of certain inherited traits from randomly mating with each other since this would result in a homogeneous distribution of gene frequencies. In human populations the mechanisms for achieving genetic isolation are geographical, social, or a combination of the two.

Geographical considerations are important because populations isolated from each other by great distances or natural barriers such as mountain ranges, deserts, or large bodies of water enjoy only limited genetic exchange with each other. Among human populations social factors can also provide the barriers to genetic exchange necessary for the maintenance of distinctive gene frequencies in different human populations. The Australian aborigine provides an excellent example of a racial group created by forces that are primarily geographical. The isolation of their ancestral homeland, the island continent Australia, has for millennia insured that mating will take place only with fellow aborigines, thereby assuring the maintenance of characteristic gene frequencies within that population. On the other hand, the American black population represents a racial group whose gene frequencies have been maintained primarily through the social factors of segregation and custom rather than geography. It must be pointed out, however, that human and other natural racial populations do not experience total genetic isolation since there are often exchanges of genes between different racial populations. Thus there is no pure race of human beings. In fact, the overlapping variation seen when one compares different racial populations of humans represents proof of their lack of racial purity and a demonstration of their hybrid character.

In discussing the question of racial origins it is necessary to recognize the role some racial traits have played in the successful adaptation of human populations to widely differing environments. During the course of its evolution, *Homo sapiens*, like other living things, responded to its environment to some extent, at least, by biological adaptation. Many of the characteristics such as skin color and body build that we recognize as associated with particular racial populations represent adaptations evolved by humans in response to the selective pressures of widely differing environments. Since its emergence in the warm sunny climates of East Africa more than 3 million years ago, the genus *Homo* has successfully invaded and made its home in vastly different areas of the world. A key to the global success of the human species has been the ability of different populations to evolve variations in skin color, body build, and physiology best suited for survival in different environments.

The differences in skin color seen in human populations arose in response to the different intensities of sunlight striking the parts of the globe that those populations inhabited. Solar intensity is highest in the tropics and least in the extreme southern and northern latitudes. Upon examination of the distribution of skin color among human populations prior to 1492 (Brace and Montagu, 1965) and the subsequent voyages of discovery, progressively darker skins are encountered as the tropics are approached. Penetration of the skin by sunlight causes the production of vitamin D in a layer of tissue just under the skin. If this layer receives insufficient sunlight, too little vitamin D is produced. Vitamin D deficiency causes the softening of the bones associated with the deficiency disease known as rickets. Too much vitamin D results in excessive calcium uptake and deposition; bones become brittle and easily broken, and calcium is deposited in the kidneys where it may produce stones. In the early days of man's existence he did not have such cultural adaptations as vitamin D-enriched milk and sun-shielding lotions and clothing to compensate for too little or too much sun. Consequently, early man's skin was the filter that regulated his absorption of sunlight. In the northerly climates there was genetic selection favoring fair skins that efficiently passed the much smaller amounts of solar radiation which fall on the northerly latitudes. On the other hand, around the tropics selective pressures favored a heavily pigmented skin that dramatically reduced the amount of sunlight penetrating the vitamin D synthesis layer. It is, therefore, not surprising that when one examines the tropical populations of Africa and Asia and their descendants one finds the brown skins of southern Asia and the black ones of Africa, parts of southern India, and Ceylon.

In addition to skin color some of the differences in body build seen when one compares different racial populations probably also arose as adaptations to climate. Contrast the long-limbed slim bodies of some African black populations with the short-limbed, generously fleshed body types found in populations of Greenland Eskimos. Both their body types are well suited to the environments where they long maintained a genetic heritage. Cold climates favor a heavy, heat-conserving body with short extremities, while hot ones favor relatively lean bodies with long extremities. Such body types have a high surface-to-volume ratio that readily exchanges heat with the environment. It is a general rule of population biology that populations of the same species found in hot climates tend to have larger numbers of individuals with heat-dissipating, high surface-to-volume ratios than their sister populations inhabiting colder climates.

It must not be assumed that every or even most racial characteristics are now or have been significant factors in the survival of the populations marked by them. While the survival value of appropriate skin color or body build during humankind's long evolutionary history may be clear, it is not possible to attribute such survival value to lip form, hair texture, blood type, and the consistency of ear wax. The racial variations seen in these traits are probably due to the operation of two factors: genetic drift, which is simply the chance fixation of certain traits, and racial interbreeding. Both the operation of genetic drift and interbreeding between racial populations can lead to the establishment of gene frequencies which, though peculiar to certain racial populations, confer no advantage or disadvantage on the population.

Race in Society

Throughout this section we have emphasized that a race is a breeding population, a population which for reasons of geography or culture mates largely within itself. Though it is true that members of the same race tend to have more of their hereditary components in common with each other than with members of different racial populations, all members of the same race are not alike. There is enormous diversity within as well as between human racial populations. Thus the concept of race has biological meaning when it is applied to populations rather than to individuals. Any system of racial classification that focuses on the individual rather than the group is bound to result in absurd and highly artificial distinctions, such as concluding that members of the same family are members of different racial groups merely because they differ markedly in appearance. Most systems that define race in individual terms are legal systems of racial classification that have been established more with an eye to politics and social custom than to biological reality.[7]

As emphasized earlier the notion of race is statistical and

describes the characteristics of populations. It does not and cannot describe a particular individual in a population. Thus any attempt to predict the body build, blood type, or even the skin color of an individual from a knowledge of his racial population may lead to erroneous conclusions. Given the internal variation of races, it is clear that the general characteristics of the populations tell little about the specifics of an individual in that population. Recently there has been considerable interest in the question of whether some racial groups have higher frequencies of genes for certain behavioral patterns than others. The answer to that question depends on the type of behavior under consideration. If it is the type of behavior which depends on traits whose gene frequencies vary from one racial population to another, then it is true that some populations may have higher frequencies of genes determining performance along such behavioral lines. For example, average population height is a race-related, genetically determined trait that is important in the type of behavior known as basketball playing, an activity that places an extraordinary premium on height. Taller populations, such as certain Caucasoid and black groups, have a distinct genetic advantage for the practice of this behavioral pattern over shorter groups such as many Mongoloid population. On the other hand, most types of behavior do not lend themselves to such a simple analysis of race relatedness. There are many behavioral patterns which, though having substantial genetic contributions, depend upon a suitable environment for their fullest development. When one attempts to make interracial comparisons of such environment-sensitive traits between racial groups whose experience and environment differ, great difficulty is encountered in deciding to what degree observed differences between racial populations is caused by racial determinants (genes) and how much is a result of the environment and hence a reflection of culture and experience.

Performance on IQ tests, for example, is a type of behavior shaped by both heredity and environment.[8] In the United States interracial differences in performance on IQ tests are matters of documented fact; and, furthermore, black populations score lower on IQ tests than do white populations (Kennedy et al., 1963; Shuey, 1966). The study by Jensen (1969), which initiated much discussion, argues the importance of genetic factors but cautions that the distinction must always be made between the individual and the population. A recent summary of thirteen studies (Loehlin et al., 1975) shows that about half of the IQ score gap (in terms of the mean) vanishes when races are matched by broad social class categories. The question is: Are blacks of inherently lower intellectual potential than whites, or does the observed IQ difference reflect the experience or environment of these two groups? A clear answer to this question is not available because no one really knows yet what the results would be if whites and blacks shared essentially common backgrounds.[9]

At present these groups by and large live apart, attend different schools, are motivated by different influences, and honor different values.

Notes

1. See Coon (1965), Garn (1965), and Goldsby (1977) for more information on races. For a contrary view, see Montagu (1964).

2. Table 2 is a partial listing of Table 2 in Coon (1965:286). Coon's table includes three additional geographical groups and the world range for a total of sixteen blood-group characteristics.

3. Adapted from Coon (1965:264). Coon's table is based upon the finding on 141 populations by Ilse Schwidetzky (*Die neue Rassenkunde*, Stuttgart: Gustav Fischer Verlag, 1962: 63-67). Although there are reports which associate nontasters of PTC with three types of disease, Coon does not think them to be important enough to explain the world distribution of tasting.

4. Adapted from Matsunaga (1962: 282). These percentages represent the calculated frequencies of the allele for wet ear wax. The ear wax types are controlled by a pair of autosomal alleles, W and w. The genotype for wet ear wax is WW or Ww, whereas that of dry ear wax is ww (page 274). Given the sample size for each population, the standard errors are calculated and presented in the third column.

5. For more information on population genetics, see Bodmer and Cavalli-Sforza (1976), Cavalli-Sforza (1974), Cavalli-Sforza and Bodmer (1971), and Stern (1973).

6. The excellent sources on the information on evolution of mankind are Dobzhansky (1962) and Lerner and Libby (1976).

7. For a discussion on problems of racial and other minorities in society, see Ehrlich and Feldman (1977) and Rose and Rose (1965).

8. For example, Jenks et al. (1972) estimated that 45 percent of IQ variation is attributable to genetic difference, 35 percent to environmental variation, and the remaining 20 percent to covariation between genotypes and environments.

9. For a critical view on the relationship between genetics and IQ, see Bodmer and Cavalli-Sforza (1976) and Cavalli-Sforza and Bodmer (1970).

2

Search for Equity on the Planet Difference

NANCY R. HAUSERMAN

In the twentieth century, a crazed scientist née philosopher combined certain chemicals (preservatives, food colorings and dyes, nuclear wastes and the like), placed the chemicals in test tubes, stored them in space capsules, and fired the rockets into the pink-hazed heavens. The space vessels eventually collided with the planet Sameness. The impact of the collision smashed the capsules and the test tubes. The result of the exposure of the stored chemicals to the atmosphere of Sameness was the creation of the species *huperson*. In point of fact, 100 hupersons were created, and these same 100 hupersons continue to exist as they existed in that first moment. They do not change; they look the same. And they *all* look the same—all hupersons look alike, sound alike, and think alike. In all manner of speaking, hupersons are alike. They are a matched set, interchangeable bookends, peas in pods. They know neither ill health nor good health, for in their sameness they have no need to distinguish good and bad. There is only one manner of being, and it is neither good nor bad; it simply is. Hupersons do not know the concept of "difference," for theirs is a world of similarity. It might be said that all hupersons were created equal. And this, of course, was the goal of the crazed scientist.

You see, the scientist existed on the planet Difference and had, at one time, been a philosopher who had read the words "All men are created equal." The philosopher had tried to assign a defensible meaning to those words and had been frustrated. The philosopher turned then from philosophy to political science and took up the task again, only to discover that a study of history and politics provided more paths connected to the same road, all leading, apparently, nowhere. Eventually, the political scientist née philosopher turned again—this time to find a meaning in the realm of law. Of course, as a lawyer the philosopher mastered the art of argument but found the rules shifted from case to case and were caveated always by exception. And so the lawyer, past political scientist née philosopher, became a crazed scientist in search of equality. The scientist never saw the planet Sameness, never saw the identical huperson and never understood how on a planet Difference anything, let alone anyone, could be equal.

Two hundred years have elapsed since Thomas Jefferson penned the words "All men are created equal." The "truth" which was considered to be self-evident has since been assigned so many meanings that whatever evidenced itself to Jefferson has since been fogged in. Yet, even absent consensus as to the meaning of the words "All men are created equal," there is a sense that whatever the truth was, it now "seems violated at every turn" (Matras, 1975:4). Indeed the only truth which many consider self-evident is not the equality among men but the inequality among people.

It was because I perceived this inequality that I attempted to understand why the conjunction "and" was used to tie together "biological differences" and "social equality," implying thereby that it was possible to acknowledge such differences and yet perceive or attain social equality. Unlike the matching hupersons, human beings represent a species rich in biological differences. We are female and male, thin and heavy, beautiful and disfigured, healthy and ill, old and young, tall and short, black and white, and so on. We do not look alike, sound alike, or think alike. We are, at least biologically, as different as we are the same. I assumed that biological differences precluded any concept of equality that was not context dependent and that inequality was inherent in these differences. It seemed to m that as expressed by Stephen (1967:264), "Upon the whole, . .

what little can be truly said of equality is that as a fact human beings are not equal."[1] Or, as Bedau (1967:4) complains, that "equality is a word so wide and vague as to be by itself almost unmeaning." Still, like many others before me, I strained to find a meaning of "equality" that transcended the creation of hupersons.

I humbly began my search with the following:

bi′o·log′i·cal, adj. (1)   Of or pertaining to biology or life and living processes

dif′fer·ence, n.   (1)   State, quality, or means of being different; variation; dissimilarity; . . .

                   (2)   Distinction, as in treatment; discrimination

dif′ fer·ent, adj.   (1)   Of various or contrary nature, form, or quality; partly or totally unlike; dissimilar

                   (2)   Distinct; separate; not the same; other

                   (3)   Out of the ordinary; unusual

e·qual′ i·ty, n.   (1)   Character or condition of being equal

e′ qual, adj.   (1)   Exactly the same in measure, quantity, number, or degree; like in value, quality, status, or position.[2]

I have arrived (in a manner of speaking, since this search is a journey with no constant conclusion) at a concept of equality that implies that all persons are entitled to equal (identical) respect for their personage. This means (Rees, 1971:102) "a commitment to respect other human beings irrespective of their abilities and achievements." It means equal consideration in the sense of equal respect for each person's life, liberty, and security. Further, as Rees (1967:122) suggests, such a concept must include the notion of "equal satisfaction of 'basic needs.' "

If we then conceive of equality as connoting a sense of equal respect for all persons, it is clear that the biological differences of human beings should have no effect on the measure of equality. "To say that men are equal is not to say they are identical" (Thomson, 1949:3). All people "are to be treated as equals, not because they are equal in any respect, but simply because they are human" (Flathman, 1967:56). It is a concept of equality which the Stoics urged—a notion that all people are equal "by nature as distinguished from convention" (Harris, 1960:4). It assumes Cicero's writing in *Delegibus* that "There is no difference in kind between man and man" (Harris, 1960:5). Equal respect is awarded solely on the basis of one's identification as a member of the human species and, as such, is determined by our "sameness"—our humanity—and not determined by our differences.

It is my belief that if society's operation were based on this ideal of equal respect, many of the present injustices would not occur. That is, if we equally consider or weigh each person's life, liberty, and security, we could not foster policies which deny not only the exercise of these interests but the recognition that the interests exist at all. This belief is postulated as an ideal. This writer recognizes the historical evidence of a dichotomy or, worse, the hypocrisy between philosophical ideals and the reality of practice. In American history, state constitutions often included "high sounding statements of the philosophy of equality side by side with property and religious qualifications (Encyclopedia, 1931:572), and so the legal and political extension of equality was limited to "free men" (Harris, 1960:18). According to Pole (1978:333), "Americans wanted a society run on equalitarian principles without wanting a society of equals."

I do not mean to imply that all persons will receive identical treatment. Identical treatment is mandated only insofar as "treat ment" refers to consideration of interests. The idea of equality which I posit herein does not insist that, as a consequence of equal respect, all persons must in all manners be treated alike. According to Bedau (1967:10), the difference may be in distribution (allotment) but not in treatment (consideration). "Equality means that men are alike only in one important specified respect, their individuality, and not that they can be treated in the same way" (Thomson, 1949:4) In fact, the notion of equal respect allows and may even require different treatment in certain circumstances. For Benn (1967:75): "Negroes and whites are equal in the sense that their interests deserve equal consideration; they are painfully unequal in the sense that society imposes on the Negro special disabilities. Equal protection ought not to mean an equal allocation of the means of protection—for the protection must be commensurate with the threat or impediment." Similarly, Sokol (1967:110) concludes that "society did not say that the presumption of equality (absolute equality) was conclusive. It simply said that unless some reason could be given for treating individuals unequally, they would be treated as if they were equal, i.e., identical."

The crucial issue is, of course, under what circumstances is such different treatment necessary or tolerable. If we acknowledge that all persons will not have the same basic needs nor the same interest in life, liberty, and security, then it follows quite logically that to meet these basic needs will often require different policies or treatments.

When I continue and say that biological differences would not affect equality, I am regarding these differences simply as characteristics which are not alike. I imply no qualitative measure. By biological differences, I mean differences like sex, color, or age which, since Rousseau, have been "thought of as 'natural'" (Bedau, 1967:10). I mean those natural or physical differences which, absent scientific intervention, are not alterable (see Rees, 1971:14).

This nonvalue-laden use of the term "difference" is crucial because it is my contention that inequality is the result of the association of values with differences. In our society, difference means more than not alike; it means as well that some characteristics (differences) are better or more highly regarded than others. We imply various values for various characteristics. The hupersons, as exactly identical creatures, would have exactly identical values. Theirs would be a society of shared values. Conversely, human beings possess nonshared as well as shared values. The young, white, healthy, handsome male possesses characteristics generally regarded as being highly valued. The old, ill, unattractive, nonwhite female possesses biological characteristics not highly valued in our society.

But to say that we value various characteristics differently does not state the entire problem. Such differential valuation is not necessarily inconsistent with my concept of equality. Rather, the essence of the problem lies in the transference of such characteristic values to consideration of the human being as a whole. Society infers that the values associated with an individual's biological differences are indicative of the individual's overall personal values. As a consequence of this transference of values, we assume that the person with the most highly valued biological characteristics is to be more highly valued as an individual. Value in one sense (biological value) becomes value in all senses.

Society makes a second assumption based on this valuation. If an individual, based on his/her biological characteristics, is assumed to be a highly valued individual, then we further assume that such person is entitled to greater respect than a less highly valued person. So, the young, white, healthy male is entitled, we assume, to more respect (respect for his basic needs and his life, liberty, and security) than is the old, ill, nonwhite female. The result is that initially we do not give each and every person the same respect. In effect, we allot respect based on the differences between human beings and not upon their sameness. When the needs and wishes of some people receive more respect and attention than the needs and wishes of others the result is inequality. For Plamenatz (1967:81):

> A difference in right or ability between two persons
> is an inequality if it gives one of them authority
> over the other, or enables him to influence the other
> much more than the other can influence him, or
> attracts more deference generally, or puts him in a
> better position to realize the aspirations he already
> has or to acquire others he should have and realize
> if he is to live well.[3]

We appear to conceive of consideration or respect as a finite resource. That is, we all start out with an entitlement to the same respect, but some people, by virtue of their biological values, are accorded a larger amount of respect. We seem to further imply that if we give more to some we must take away from others. It is as if equality, or more particularly the respect implied therein, were a pie. We begin with an assumption that we are entitled to an identical share of the pie. Obviously, if we give one person a larger piece than his/her original share (original meaning as if in the beginning each human being had pieces of identical size) we must decrease the share of another person.[4]

Furthermore, when, as a consequence of giving some people larger shares, society feels compelled to reduce the shares of others, there must be a consideration of whose share to reduce. Again, the biological differences determine the share—those less valued persons receive the reduced portions of the pie.

The result of the redistribution is unequal consideration. The result of the unequal consideration is social inequality.[5] Matras (1975:9) for instance defines social inequality as "that inequality in the distribution of social rewards, resources, and benefits, honor and esteem, rights and privileges, and power and influence which is associated with differences in social position." Again, I do not mean to imply that social equality would necessarily result in all persons having identical rights, privileges, and benefits. The inequality of distribution means that basic needs and interests in life, liberty, and security were not equally regarded for all persons.

Social inequality is a state of affairs which perpetuates itself since, according to Matras (1975:6), "It is partly *because* of their [social positions'] different institutionalized claims and access to social rewards and resources that the different social positions are differently valued and attractive." So, if as a result of high value based on certain biological characteristics, a person receives a greater share of respect than any other person, she/he is likely to have access to more social rewards. This person then has high value based on biological characteristics and high value from social position. The consequence is that the overall value of the person continues to increase, resulting in still a greater share of respect, therefore greater social access, and so on.

Of course, the corollary to some people getting greater shares of respect and hence increased access to social rewards is that those people who initially receive less respect based on the low value of their biological characteristics have less access to social rewards and valued social positions. Consequently, these people remain of low value and their basic needs and interests continue to receive less attention.

Having begun with a concept of equality that excludes biological differences as a determinant, there is an obvious conflict between the ideal of equality and the operation of a reality in which such biological differences clearly do make a difference. Stated another way (Wilson, 1966:8), it may be that "similarities and differences such as color, facial features, shape of skull and so on are not relevant to the way men ought to be *treated.*" But, as Wilson goes on to point out, relevance depends on what we count as relevant, and it is apparent to me that since biological differences do affect our treatment of others, they are (aside from whether they "ought" to be) relevant.

In the midst of these differences, how can we assert, or reassert, a concept of equality? The ideal solution would be to change society's perception of equality. This infers a reeducation to the effect that we accord each person an equal amount of respect and do not vary this respect based on biological differences. It is, of course, no easy task to change the attitude of any society. It is an especially difficult task if we concentrate our efforts solely on the attitude or philosophy which lies behind people's actions. Instead, we can attempt to change or regulate external behavior patterns in the hope that such change will, eventually, result in the attitudinal change. In this country, such policy changes are both affected and effected by the role of the courts and the government. Courts and legislators are both the innovators and the followers. They cannot help but be influenced by the social prejudices and the social objectives of the culture in which they live. At the same time, legislative enactments may reflect the ideals of a society, and court review of such legislation presumably ensures that such laws will be free from inequalities based on prejudice.

Such policy changes are not necessarily attempts to achieve equality, however. Rather, policies or programs may be attempts to remove, or at least diminish, inequality. They may be attempts to change what exists and not necessarily attempts to create anew (Pole, 1978:358).

One such external change is an attempt to eliminate biological differences. Throughout the course of the summer institute on biological differences and social equality, I was confused by the paucity of philosophical discussion. The lectures and discussions did not feature concepts of equality but were more often about various biological or quasibiological subjects such as sickle-cell anemia, genetic screening, evolution, and IQ. It now seems quite clear to me that recognition that biological differences do affect equality has led some geneticists and others to regard the elimination of such differences as crucial to the issue of inequality. Hence the discussions about differences in IQ, how to raise IQ, sickle-cell anemia, and human evolution and the like.

As noted by Rees (1971:102), however, the "very fact that 'genetic' engineering should be envisaged at all as a possible remedy for inequality of opportunity suggests such a strong desire to ensure uniform treatment for all persons as to exclude or belittle other values." The creation of such a society (much like the hupersons) would destroy that element of equality which implies "a commitment to respect other human beings irrespective of their abilities and achievements" (Rees, 1971: 102). Indeed, equality is "legitimized by a profound, not merely perfunctory, respect for individuality. . . . [it] emphasizes the distinctions among people as well as their similarities" (Pole, 1978:x). I do not mean to imply that all corrective devices or the elimination of disease are undesirable goals. I mean only that we need not, and must not, eliminate the uniqueness of each human being to achieve equality.

In addition to genetic engineering and an alternative methodology when such biological change is impossible, policy makers seek to effect equality by reslicing the pie. In other words, programs are designed to provide the advantages which we assume would have accrued to some people except that they were accorded less respect based on their low valued biological characteristics. It is this type of solution which results in, for example, affirmative action programs. We look at a situation which appears unjust and assume that such situation occurs because, initially, all persons' basic needs and interests in life, liberty, and security were not equally considered. Again, this does *not* mean that programs such as affirmative action treat all persons equally. Rather, such programs are often attempts to rectify or lessen the injustices which result from the inequality of respect shown. "When egalitarianism is translated into concrete political programs," according to Benn (1967:64), "it usually amounts to a proposal to abandon existing inequalities, rather than to adopt some positive principles of social justice."

Though such programs cannot force people to show more respect to those persons of "less value," the programs try to alleviate the impact of such differential respect. Needless to say, when the pie is redistributed some who had more will now have less than they previously had, although their portion may still be considerably higher than most. Those facing this reduction in share will necessarily view this reslicing as unjust and unequal treatment.[6] "Perhaps not all inequalities can be rectified; and it is certain that some can be rectified only by creating new inequalities and new grievances" (Pole, 1978:326).

Conclusion

As the concept is discussed in this paper, equality is descriptive, evaluative, and distributive. It implies an equal (same, identical) amount of respect to each person based solely on her/his humanity. Respect connotes consideration of each person's basic needs and her/his life, liberty, and security. In this way, equality stresses the sameness (humanity) of society while not denying "legal and moral individualism." Once one is established as a member of the human species, differences *should* have no effect on equality (equal respect), and we could speak of biological differences *and* social equality without appearing inconsistent.

As noted above, equality does not imply value nor is it value-dependent. However highly regarded certain biological characteristics may be (e.g., health, youth, beauty), we should not confuse whatever value judgments we may make about these various biological differences with the measure of respect accorded to each person's basic needs and fundamental interests. To imply that some people are more deserving of this basic consideration is to promote inequality. The inequality of consideration results in social inequality.

If social inequality is an "artificial creation of men" (Theibar and Feldman, 1972), then it can be eliminated, naturally or artificially, by these same people. Inequality is a condition that could be manipulated and that human energy could change. I cannot agree that inequality is a necessary feature of any social system. Such would be true only if we conceive of any nonidentical treatment, right, or privilege as being an inequality. In the context of this paper, inequality means only that we do not treat people alike in one important aspect—respect based on humanity.

Equality does not mean that all people will have all things. Clearly there are many valid and relevant factors affecting distribution of societal rewards and benefits. The distinction, which is in part the point of this paper, is that there are no valid or relevant factors which should affect the consideration of each person as a human being. In this manner, we all at least begin from the same starting point. Biological differences should not be used to handicap some persons and give large leads to others. In this way we can begin as offered by Wilson (1966:64) "to arrange things so that merit and desert are rewarded, rather than qualifications which the individual can do nothing about."[7] Benn (1967:69) argues "Whereas we respect different men for different things, there is no property, such as white skin, which is a necessary condition of a man's being worthy, whatever his other merits, of any respect at all. So every man is entitled to be taken on his own merits; there is no generally disqualifying condition."

In the end there is perhaps no one, correct, and ultimate answer to the riddle of defining equality. It seems clear that equality is not an end in itself and perhaps should be considered only as a beginning or starting point. It is no easy task to determine the path which one should follow with equality as the point of embarkation. The movement towards social equality is a "movement towards a vision but without a destination. To appear to have arrived at the answer is only to have reached the point where the problem has changed" (Pole, 1978:355).

Notes

1. Stephen's pessimism is shared by Sokol (1967:91), who writes that "As a philosophic concept the doctrine or dogma of equality was essentially meaningless. As a philosophic tool, it was useless."

2. *Webster's New Collegiate Dictionary*, 1949 ed.

3. I have inferred that "natural" rights are implicit in the use of the word "right" in this quote from Plamenatz.

4. Such a quantitative concept of equality would seem to be a faulty one. Unlike the circumference of a pie, there is no absolute boundary to respect. In the ideal, respect available for distribution is limitless. Yet while this limitlessness may be the reality, it is not the reality by which our society appears to operate.

5. Concepts of social equality are often only implicitly expressed as corollaries to discussions of social inequality. Social equality as such is not defined, but the reader may infer that the opposite or elimination of social inequality would produce, conceptually, social equality. It is also important to distinguish between human equality and social equality. See, e.g., Catlen (1967).

6. The "reverse discrimination" suits recently before the U.S. Supreme Court are illustrative of persons who feel that their "share of the pie" was unjustly reduced, and, apparently equally as important to them, the shares of others were increased without regard to actual merit.

7. As a caveat, by the use of this quote I do not mean to imply my advocacy of a meritocracy, a political philosophy with which I have some reservations.

3

Evolution, Ethics, and Equality

STEPHEN L. ZEGURA,
STUART C. GILMAN, AND
ROBERT L. SIMON

Introduction

Darwin's theory of natural selection still forms the cornerstone of the modern synthetic theory of evolution. The interaction of this biologically based paradigm with theories of social change has led to two conspicuous attempts to explain (and in some instances, justify) social phenomena. These intellectual movements are social Darwinism and sociobiology. Social Darwinism is probably a misnomer because Herbert Spencer formulated many of his central ideas before the publication of *On the Origin of Species* in 1859 (Harris, 1968). In this sense Spencer cannot be accused of simply applying Darwin's ideas to societal issues. Sociobiology, on the other hand, does represent the direct application of Darwinian ideas to the study of social behavior.

The search for the biological bases of social behavior has a long history involving the development of disciplines such as ethology, ecology, population biology, and comparative psychology. In 1975 E. O. Wilson presented a synthesis of these various approaches within the framework of evolutionary theory. His book *Sociobiology* has been strongly criticized by social scientists (and by many biologists as well) for its lack of caution when extrapolating to human beings. More recently Wilson has attempted to answer his critics by writing a book which specifically applies sociobiological principles to the human condition (Wilson, 1978). Once again, Wilson has been accused of naive reductionism, unfounded analogy, biological determinism, unsophisticated Marxism, and numerous other scholarly sins (Gould, 1978). Sociobiology, however, is not likely to disappear because it is unpopular in some circles. Investigation of its theoretical foundations and potential applicability seems a more worthwhile approach than throwing stones. It is to a small subset of these concerns that this paper is addressed. Collaboration of a physical anthropologist, a political scientist, and a philosopher interested in ethics hopefully will prove synergistic for the explication of how ideas from evolutionary biology, social Darwinism, and sociobiology might relate to ethical theory and the issue of equality.

Evolutionary Theory

Evolutionary theory provides an explanatory framework for the phenomenon of biological change through time. Darwin considered evolution to be "descent with modification." Modern neo-Darwinians have a variety of definitions for evolution. These definitions cluster into two groups which emphasize either gene frequency change or adaptation as the key criterion for demonstrating evolutionary change. For instance, in discussions of evolution by population geneticists the most commonly encountered definition of evolution is "change in gene frequency through time." Evolutionary biologists such as Ernst Mayr with a background in systematics often stress adaptation in their definitions. Mayr's most recent definition of evolution is "change in the diversity and adaptation of populations of organisms" (Mayr, 1978). Most biologists concur that change through time, whether at the level of genes or morphological and behavioral adaptations, is a demonstrable reality. Where opinions differ involves the explanation of these changes, the subject of evolutionary theory.

The modern synthetic theory of evolution represents the outcome of a merging of disciplines as diverse as population genetics, comparative anatomy, population biology, evolutionary biology, systematics, paleontology, and ecology. In discussing the four forces that directly produce evolutionary change it will be helpful to employ a consistent point of reference. In this context change in gene frequency represents the fundamental descriptive parameter of evolution. The four forces or processes that can lead to changes in gene frequencies are mutation, natural selection, gene flow (migration), and genetic drift.

Mutation is the ultimate source of all genetic variation. It is a necessary antecedent to the genetic processes of meiotic crossing-over (recombination) and independent assortment which increase genetic variability by reshuffling previously existing genetic variants. Mutation is defined as "any heritable change in the genetic material." At the biochemical level this involves changes in the sequence of the information-carrying portion of the hereditary material, deoxyribonucleic acid (DNA). In humans mutation rates per gamete generally range from 10^{-6} to 10^{-4} for a particular gene. Most of these mutations are harmful to their possessor. The highest mutation rate recorded in our species is 2×10^{-4} for neurofibromatosis, a disorder which affects the nervous system and pigmentation of the skin (Dobzhansky et al., 1977). The biochemical correlates of many mutations are known. For instance, sickle-cell hemoglobin, which is responsible for sickle-cell anemia, can be traced to the substitution of a single amino acid (the chemical units which are combined to produce proteins). Further analysis shows that this amino acid substitution is caused by the smallest possible alteration of the genetic message encoded in the DNA responsible for the production of the hemoglobin molecule. Many other mutations can also be traced to minor changes in the hereditary material. However, there is another class of mutations called chromosomal aberrations that involves larger genetic units. These errors include the addition and deletion of whole chromosomes (as in Down's syndrome and Turner's syndrome in humans), the rearrangement of the spatial positions of genes on chromosomes, and the addition or deletion of one or more genes on a chromosome. Together the small point mutations and the larger chromosomal aberrations provide the raw material for evolutionary change because a small portion of these mutations turn out to be beneficial for their possessors.

The process by which the raw material for evolution is sifted or molded is called natural selection. Natural selection is generally considered to be the most important evolutionary force. Darwin and Wallace first developed the theory that natural selection was responsible for evolutionary change. Darwin defined natural selection as "the preservation of individual differences and variations and the destruction of those that are injurious" (Darwin, 1859). For Darwin differential survival and differential mortality were the most important components of the process of natural selection. Modern neo-Darwinians have altered this focus to highlight differential reproduction as the central component of natural selection. As a result the modern definition of natural selection is "differential reproduction of genotypes." In other words, the possessors of certain genes leave more offspring than the possessors of alternative genetic instructions. A person's genotype refers to the entire genetic constitution of the individual. The term "genotype" can also be used to refer to the genetic constitution of a person at a single gene locus (the spatial location of a specific gene on a particular chromosome). Both differential mortality and differential reproduction influence the process of natural selection. In the human species culture and technology interact with our biology to produce situations in which these two components are of differing import at different times in different environments. For instance, Western medical technology has greatly reduced mortality in modern industrial nations, thereby decreasing the opportunity for natural selection due to differential mortality. The combination of mutation and natural selection still represents the principal means by which evolutionary change takes place according to the neo-Darwinian synthesis.

The third evolutionary force is called gene flow or migration. It is defined as "the exchange of genetic information between different populations." This exchange often results in gene frequency changes in both the donor and recipient populations. Sometimes the process involves temporary contact during which reproduction takes place. In other instances a more permanent spatial relocation of organisms is implied. Like natural selection, migration has a directional component in contrast to mutation and genetic drift which are random processes. Modern technology has influenced the importance of migration as an evolutionary force in our species. The ability of people to move about has dramatically increased over the last few hundred years. This increased mobility is reflected in the increased exchange of genetic material between previously isolated populations.

The fourth and final evolutionary force is genetic drift. Genetic drift refers to "random fluctuations in gene frequencies due to chance occurrences." The effects of genetic drift become more important in small, isolated populations. In large populations natural selection frequently counters and swamps gene frequency change due to genetic drift. The major cause of genetic drift is sampling error at fertilization. By chance, genes may be present in the subsequent generation at frequencies that do not correspond to their frequencies in the previous generation. The phenomenon known as the founder effect is also a part of genetic drift. If a small portion of a population leaves the original population to form a new breeding unit, the gene frequencies of this founder group may differ from the gene frequencies of the parent group because of sampling error. Modern technology has reduced the number of isolated groups in our species, and the opportunity for genetic drift in human populations has likewise declined. The current emphasis on understanding evolution at the molecular (as well as organismic) level, however, has supplied a vast amount of evidence concerning protein structure that points to a significant role for genetic drift in the overall process of evolution.

The most important point to remember about the four forces of evolution is that they usually operate simultaneously in any particular situation. Sometimes they act to enhance one another's effects on gene frequencies. In other cases they effectively cancel one another out. It is the total interaction of these four processes that produces evolutionary change.

The Unit of Evolutionary Change

There is great debate over the fundamental unit of evolutionary change among modern biologists. Many evolutionary biologists claim that only populations evolve. Individuals possess genes but it is the collectivity called the gene pool of the population which actually changes through time. There is also disagreement concerning what level of population represents the actual evolving unit. Biological populations represent interbreeding groups of organisms, but there are different levels of these interbreeding units called Mendelian populations. The most inclusive level of Mendelian population is the species, group of actually or potentially interbreeding natural populations which are reproductively isolated from other such groups (Simpson, 1961). The least inclusive Mendelian population is the deme or local population within which mating usually takes place. Both the species and the deme are considered to be the actual units of evolution by different authors (Dobzhansky et al., 1977; Mayr, 1978). Other authors point to the fact that natural selection operates at the level of the individual organism, and, therefore, individuals, and not populations, are the units subject to Darwinian evolution (May, 1978). Some sociobiologists carry the process of reductionism even further and state that the individual gene is the fundamental unit of evolution (Dawkins, 1976). In this conceptualization organisms are survival machines run by a set of genetic instructions for the "purpose" of perpetuating genetic entities.

If one adopts the definition of evolution—change in gene frequency through time—Mendelian populations are logical choices for the units of evolution because gene frequencies are properties of populations. In some cases the species may be the best candidate for the evolutionary unit (as in the gradual transformation of a population through long time periods). In other situations where rapid evolutionary change involves the branching off of a small peripheral deme that subsequently acquires reproductive isolating mechanisms preventing interbreeding with other demes of the parent species, the evolutionary unit may well be the deme. In either case an important component of the evolutionary process is the genetic structure of these populations.

Population Structure

Ultimately population structure involves which gametes unite with other gametes to form the gene pool of the next generation. Mate selection and all those factors which influence mate selection are integral parts of population structure. Geographical distance, courtship behavior, individual preference, mate availability, and, in human, sociocultural proscriptions all interact to determine which gametes finally unite.

The simplest mating system is called random mating. Random mating refers to the situation in which any individual of one sex an equal opportunity to mate with any conspecific individual of the opposite sex. If one analyzes our species, this theoretical equality of opportunity is seldom obtained. In fact, people usually end up mating with individuals who are similar in certain observable traits. Tall men frequently marry tall women. People with high IQs usually marry people with similar capabilities, and people often marry preferentially in terms of skin color. These examples are all deviations from a random mating pattern. They involve the phenomenon known as positive assortative mating— the preferential mating of phenotypically similar individuals. In other words, people who are similar in some observable characteristic (the phenotype) mate more frequently than would be expected by chance given the frequencies of the particular traits in the population.

Humans have also developed elaborate sets of cultural rules for marriage embodied in intricate kinship systems. These rules often limit potential mates to particular categories of kin. Not only are we sometimes told whom we can marry, we are often told whom we cannot marry. Culture can thus foster another kind of deviation from random mating. This phenomenon is known as inbreeding. Inbreeding refers to the mating of related individuals more frequently than would be expected by chance. For example, if the kinship rules of a society state that a male should marry his mother's brother's daughter, then the frequency of marriages between first cousins will be far greater than the product of the relative frequencies of male and female first cousins in the population.

These two deviations from random mating (inbreeding and positive assortative mating) have important genetic consequences in terms of evolutionary change. Although they do not directly change gene frequencies like the four forces of evolution, they can, in certain circumstances, indirectly affect the evolutionary process. Specifically, they both lead to an increase in the population of the proportion of homozygous genotypes (the situation in which a person receives the same genetic message at a particular gene locus from both parents). This can prime the population for natural selection to eliminate deleterious genes more effectively. Many potentially harmful recessive genes are masked when they occur in combination with a dominant gene (the heterozygous genotype). Inbreeding and positive assortative mating increase homozygous combinations at the expense of heterozygous combinations. As a result, heterozygous combinations decrease in frequency, thereby decreasing the number of protected recessive genes. So just as the four forces of evolution interact with one another, they also can interact with elements of population structure to produce evolutionary change.

Challenges to the Neo-Darwinian Perspective

The modern synthetic theory of evolution stresses mutation and natural selection as the prime movers of evolutionary change. Migration and genetic drift are relegated to subsidiary roles. Ever since the appearance in 1969 of an article by King and Jukes entitled "Non-Darwinian Evolution," the field of evolutionary theory has been involved in heated debate. King and Jukes contend that most evolutionary change in proteins may be due to mutations that are effectively neutral with respect to natural selection. They also propose that many genes responsible for these proteins exhibit frequencies that are the result of genetic drift acting on these neutral mutations. Nearly a decade of observations and experimentation has not produced a definitive answer to the question of the relative import of natural selection and genetic drift in the explanation of genetic variation. Both evolutionary forces are probably operating, and in some situations natural selection may be the more autocratic dictator of evolutionary change, while in other instances genetic drift may be the stronger force.

Sociobiologists generally take the position that natural selection is of paramount importance in explaining the evolution of behavioral patterns. They are, in effect, extending the application of natural selection as the explanatory paradigm of genetic and morphological change to the realm of behavior. The basic assumption that sociobiology makes is that there is a biological basis for social behavior. Specifically, behavioral capabilities have their roots in the genetic constitutions of individuals. Some sociobiologists claim that particular genes or sets of genes predispose individuals to specific behaviors. Others are more cautious because of the complex interaction of the genotype and the environment which together produce observed behavior. Sociobiology provides a very different kind of challenge to neo-Darwinian evolutionary theory. For our species it entails the application of biologically based ideas to realms of inquiry previously considered to be the province of the social sciences. Like social Darwinism before it, sociobiology has been used as a justification for position statements regarding social issues. Social class, social change, equality and ethics are some of these social issues which will be discussed in terms of the framework provided by evolutionary theory. We intend to do this by first examining the historical development of social Darwinism, after which we will return to the topic of sociobiology.

Historical Overview of Social Darwinism

In order to examine the relationships among evolutionary theory, sociology, and social Darwinism it is important to view the unfolding of the latter as a dominant paradigm in the social sciences during the late nineteenth century. This might appear to be far removed from the interests of the biological scientists articulated in the past several pages; however, both historically and epistemologically there has been a close relationship between social ideas about evolution and biological ideas about evolution. Using the previous discussion as background, we will now turn to a historical overview of this relationship.

One of the difficulties in trying to understand the changes wrought by new paradigms like Darwinism is the "nominal" assumption that Charles Darwin was responsible for the movement. It would perhaps be easier to understand these great changes if it were first understood that Darwin was, in a sense, the great synthesizer of what we now refer to as Darwinism. Darwinism was actually a culmination of the work of William

Wells, Charles Lyell, Erasmus Darwin (Charles's grandfather), Jean de Lamarck, Thomas Malthus, as well as many others (Barzun, 1958; Greene, 1959). Darwin effectively summarized earlier ideas in his book *On the Origin of Species*, adding the first clearly documented exposition of natural selection. This magnum opus has a curious history. After making it aboard the H.M.S. *Beagle*, Darwin was not really certain of the importance of his work, was disturbed and embarrassed by the awkward language in the page proofs, and was constantly waivering over whether he really believed in natural selection. Yet, because of chance and persistence, Darwin became one of the giants of modern intellectual history.

It is also interesting to note that if it were not for some strange quirks of Victorian society and his own personal social relationships, Darwinism would now probably be known as "Wallacism." Given contemporary accounts, Alfred Russel Wallace may have been a better scientist than Charles Darwin. Wallace and Darwin had conceived the same idea almost simultaneously. But Darwin published a book, whereas Wallace only published a scientific paper. This would not have been such a crucial difference had Darwin's book not fitted the mold of "popular science" during the late nineteenth century and had his work not been positively reviewed in the *Times* by the influential T. H. Huxley (Barzun, 1958). Even more importantly, Darwin's work could be talked about over tea without an in-depth reading for the kernel of the book was brief while its proof comprised the vast majority of the text.

All of this is telling if we are to gain some understanding of Darwin's social intent. For until he published *On the Origin of Species*, he was noted for the obscure nature of his work and for being Erasmus Darwin's grandson. In a very real way Darwin was thrust into the spotlight of world renown, and he was poorly prepared for this role. Most historians of science see Darwin as a competent observer, but not all see him as an intellectual giant. This is a fascinating description which differs from the public perception of Darwin. He did rather poorly in school, and, pushed on more by family heritage than ambition, he finally attended the medical school at the University of Edinburgh. While there he made a close friend of Dr. Grant, an ardent Lamarckian, who greatly affected Darwin intellectually. Thus, when Darwin found himself thrust among the luminaries of England, he was not only outclassed, but struck by their imaginative extrapolations from *his* theory. For example, he was amazed how creative individuals were to make such profound observations as, "during the late ages, the mind will have been modified more than the body; yet I had got so far as to see with you that the struggle between the races of man depends entirely upon intellectual and moral qualities" (Barzun, 1958; cf. Goudge, 1961). Darwin readily admits that he found Wallace, Spencer, and Huxley indispensable sources of ideas.

Although it is unfair to suggest that a person should be known by the company he keeps, it is obvious that Spencer and Huxley had a major role in the ultimate impact of Darwin's work. Even before the publication of *On the Origin of Species*, Darwin had been significantly influenced by Thomas Malthus, who had also affected most of Darwin's contemporaries. "It is the doctrine of Malthus," writes Darwin in 1844, "applied in most cases with tenfold force. . . . Yearly more are bred than can survive; the smallest grain in the balance, in the long run, must tell on which death shall fall, and which shall survive. Let this work of selection, on the one hand, and death on the other, go on for a thousand generations; who would pretend to affirm that it would produce no effect" (quoted in Greene, 1959:262). Additionally, in this treatise Darwin makes note of the impact of racial differences while using Malthus's notion of "struggle for life" as his ascriptive metaphor. Therefore, although social implications are buried in Darwin's work they can be exhumed and additional support can be developed by noting those individuals he looked to for intellectual nourishment. Darwin was a nineteenth-century Victorian belabored with all of the notions of manifest destiny and racial (and class) superiority. He was also willing to see his work extended to those seemingly implied conclusions.

However, it would be an error to dismiss Darwin's contributions on these grounds alone. He did provide one of the most important paradigms in modern science in *On the Origin of Species*. Major difficulties arise with his theory only when they are pushed beyond their biological context into the social and political arena. Admittedly, there is no firm borderline between these areas, but, as we shall see, many of the critiques of Darwin's biological theories (and subsequent issues in sociobiology) stem from these problems. Many philosophers believe that the theory of evolution is not a theory; however, David Hull has emphasized that most of these criticisms are based upon a weak foundation in biology and gross overstatements by the critics (Hull, 1969). But, even more important, as Michael Scriven pointed out two decades ago, we cannot dismiss evolutionary theory because it has not developed the ability to predict specific events (Scriven, 1959). The weaknesses of the Darwinian paradigm involve two issues which are fundamentally ethical.

The first difficulty arises in the process of definition in scientific language. This has been noted by many others in the context of sociobiology (Sahlins, 1976), but few have pointed out that most (if not all) of these problems arise in the application of such definitions in biology. For instance, the definition and meaning of "fitness" has been at issue for the past 100 years. Many have argued, and we think wrongly, that "fitness" itself is a tautology (Bunge, 1961; Flew, 1967; Goudge, 1961). The use of the idea of fitness is relative. "Certain organisms in a given environment are fitter than others. A higher percentage of those organisms which are nearer the 'fittest' end of the scale tend to survive than those at the other end. This scale in turn is ordered at least in part independently of the actual survival of these individuals. Of course, the claim that the fittest tend to survive can be made viciously circular if fitness were determined only by means of actual survival or into a tautology by defining 'fitness' exclusively in terms of actual survival, but biologists do neither" (Hull, 1969:243).

The problem with the notion of fitness is that it exists within our language structure as two simultaneous ideas. The first part involves the biochemical and physical components of fitness. The second element of the definition includes a notion of history, that is, a developmental perspective. Unfortunately, this second element necessarily includes, and logically appears to force, social conclusions from biological functions. Without history it is impossible to accept any notion of evolution.[1] However, Darwin appears to have "two minds" on this issue.

In his purely biological work, he suggests that history is devoid of social content, and selection is both blind and nonrational. That is, no human judgment can determine what is fit except after the fact. This emphasis on "postdiction" rather than "prediction" does not make such an enterprise nonscientific, just different (Scriven, 1959). But Darwin, especially in works like *The Descent of Man and Selection in Relation to Sex* (1871) and *The Expression of the Emotions in Man and Animals* (1872), also implies a social dimension of fitness which is at least disturbing. For if nature is seen as a mentor, it is only a small step to suggest that fitness and natural selection are "a sort of supreme court of normative appeal" (Flew, 1967:17).

The second major ethical difficulty is found in the utility of observation for supporting evolutionary theory. Certainly *On the Origin of Species* demonstrates Darwin's ability to observe with a "keen eye" and to synthesize this material into a persuasive theory. Darwin also used his observational abilities, firmly embedded in his cultural milieu, to extrapolate social conclusions which even his most ardent defenders would find hard to justify (cf. Hull, 1969). Although several instances have been mentioned previously, perhaps an elaboration of one example would be illustrative of the ethical issues entailed.

Darwin apparently believed that insanity was simply the expression of an earlier evolutionary stage and that the insane were interesting examples for demonstrating an aspect of evolutionary theory (Gilman, 1979). In order to study the possible relationship between insanity and evolution he employed the work of James Chrichton Browne, a distinguished nineteenth-century psychiatrist who was the medical director of the West Riding Asylum. Browne was also an amateur photographer, and Darwin became enthralled with his photographs as demonstrations of "human development." Darwin viewed pictures of three cases of hypochondria "in which the grief-muscles were persistently contracted." He also noted the "angry scowl" and the "prominent row of hideous fangs" of an epileptic idiot and became fascinated with the phenomenon of "bristling hair" in several photographs, which he incorrectly assumed was associated with insanity. This suggests that Darwin was selecting arbitrary characteristics as "indicative for the entire physiognomy of the patient" (Gilman, 1979). Such a selection would allow an emphasis upon the "animal" characteristics of the insane as support for his social extrapolation of evolutionary theory.

This example demonstrates an ethical problem of evidence within evolutionary theory. The difficulty with the theory is that it is so vitally tied to history that modern examples (which cannot be observed through any meaningful period of time) cannot be disputed. For as long as an event is contemporary it can always be interpreted as evidence, and there is no way of falsifying it—one can only offer alternative explanations. This is one reason why Michael Scriven (1959) suggests that biology is an "irregular subject."

Darwin did become disillusioned with the use of photographs because he ultimately saw how arbitrary their use was (Gilman, 1979). He continued to believe that the insane did fit into evolutionary theory, but apparently was an astute enough scientist to see the limits of such an ahistorical argument. Many of his contemporaries, those now labeled social Darwinists, were far less reticent about the use of contemporary examples to demonstrate their thesis of "survival of the fittest" and made the social scope of their works even more explicit.

Spencer's sociology used the notion of evolution, ultimately joining natural selection to it in his later works, to describe and rationalize all modes of human behavior. This treatment tended to exalt the market economy and cultural characteristics such as industriousness, cleanliness, and thriftiness. For Spencer, each of these characteristics demonstrated the superiority of one race over another and acted as a justification for social class (Spencer, 1884). Although Spencer's notion of social justice is open to severe criticism, it is important to note how widely it was accepted (Miller, 1976). Nature was envisioned as a hard taskmaster, and this tended to rationalize the harshness of the social and economic circumstances of nineteenth century industrialism. Spencer saw social types like the drunkard as naturally filling out the context of social evolution. Using Malthus to build on Lamarck, he was able to develop an all-encompassing social theory. For this reason he emphasized that a fundamental understanding of biology would lead to a more sophisticated understanding of society:

All the animal functions, in common with all the higher functions, have as thus understood, their imperativeness. While recognizing the fact that in our state of transition, characterized by very imperfect adaptation of constitution of conditions, moral obligations of supreme kinds often necessitate conduct which is physically injurious.

Spencer continues,

And this conception of conduct in its ultimate form implies the conception of a nature having such conduct for its spontaneous outcome—the product of its normal activities (Spencer, 1880:87; cf. Dunn, 1965:33-38).

Although a bit crude, this certainly is a precursor to the explanations of altruism developed by sociobiologists. The difference, which will be examined shortly, depends on the degree of explanation.

The historical question which has yet to be answered is if Spencer and others were Lamarckians how is it that they were included in the "Darwinian revolution"? In any contest between two paradigms "war" is the closest description for what occurs between the two sides of the argument (Kuhn, 1970). When the forces of Darwin won, they exacted great tolls from those who were considered enemies. They also took care of their friends.

Once in power the victors could afford to manhandle past and present dissenters or applying a double standard, make all the exceptions in their own favor: allowing Lyell's reservations about man because he accepted Natural Selection; letting Wallace, for the same reason, believe in the separate origin of man's soul; overlooking Lewe's anti-Mechanism; allowing Spencer his ancient and revised Lamarckism because he joined Natural Selection to it, while Butler was boycotted; numbering

> all the American Darwinists as friends, even though
> theists, and yet keeping silence about Agassiz be-
> cause he would not depart from original creation;[2]
> overlooking Darwin's unlucky theory of Pangenesis,
> his confused expressions, and his occasional errors
> of fact . . . (Barzun, 1958:119).

Thus, social Darwinism was "piggybacked" upon Darwinism and was accepted because of the victory of the Darwinian paradigm. The social Darwinists also won because the theory gave apparent "proof" to the belief that races were biologically inferior. Francis Galton (Darwin's cousin) even provided a simple quantitative model of intelligence (much reminiscent of IQ) in which he ranked in ascending order Negroes, Englishmen, and the "Athenian race" of the fifth century B.C. (Provine, 1973).

The legitimacy of social Darwinism became manifest in the United States through the work of Yale sociologist William Graham Sumner. Sumner, like Spencer, saw social equality as not only idealistic but dangerous. "If there is any place where men are equal, it is not the cradle but the grave." In fact, he saw justice as not entailing equality of political power, equality of possession, or enjoyment of life. The status quo was by definition the ultimate and most sensible outcome of human evolution. Spencer saw the status quo not in democracy, for "democracy cannot thrive when things are hard," but in aristocracy. The biological interface was with the economy and its emphasis on capital. The problem with society was that its laws and institutions tended to work against capital "to produce a large population, sunk in misery. All poor laws and all eleemosynary institutions and expenditures have this tendency. On the contrary, all laws and institutions that give security to capital against the interests of other persons than its owners, restrict numbers while preserving the means of subsistence" (Keller, 1914:27-28). For the late nineteenth and early twentieth century, social Darwinism worked both as a champion of the business ethic and a hidden enemy in the camp of progressive democracy (McCloskey, 1964).

The ethical problem of social Darwinism is firmly rooted in

the works of Herbert Spencer. Spencer unabashedly gave a full-blown social and historical interpretation to Darwin's evolutionary theory. Where Darwin was reluctant, Spencer was petulant. At its base social Darwinism is a world view which existed in form, if not substance, in earlier writers such as Thales, Anaximenes, Telesio, Büchner, and Haeckel. The difficulty is that such theories lack scope, for they cannot rely on mere evidence. "For its adequacy depends on its potentialities of description and explanation rather than upon the accumulation of actual descriptions, though its power of description is never fully known short of actual performance" (Pepper, 1966:97). Thus, the problem with social Darwinism is that *everything* could be explained as fitting within the theory as long as some explanation could be offered. In this sense there was no way to disprove the theory because proof (short of projection and supposition) would only come over periods of thousands of years.[3]

The problem of Spencer and the social Darwinists is not that they led to a conservative movement per se. For they could have just as easily led to an extreme leftwing logic (Sahlins, 1976). For instance Friedrich Engels suggests that Darwinism "is the foundation for communism." The problem accompanying the cultural vogue of individuals like Spencer and Sumner was that they tended to set the social and political agenda for entire nations. It would be reasonable to suggest that such movements were originally conservative and simply tended to adopt and mold evolutionary theory into the kind of paradigm which most closely justified their position.

Certainly the most profound impact of social Darwinism was in the eugenics movement in the United States. Although Francis Galton can be called the father of eugenics, Louis Terman, who introduced the Stanford-Binet IQ test, is certainly the father of its application. According to Terman: "If we would preserve our state for a class of people worthy to possess it, we must prevent, as far as possible, the propagation of mental degenerates . . . the increasing spawn of degeneracy" (Terman, 1917:165). But who were these mental degenerates? Goddard used the new mental tests to examine immigrants on Ellis Island and discovered that 80 percent of all Hungarians, 83 percent of all Jews, 79 percent of all Italians, and 87 percent of all Russians were "feeble-minded" (Kamin, 1976:376-77). These data were used by eugenicists and geneticists to argue in favor of the Johnson Immigration Act of 1924 which until 1965 spectacularly limited the number of southeastern European immigrants. This law was responsible for turning down the application of hundreds of thousands of Jews between 1939 and 1945, ultimately leading most of them to death in the concentration camps.

It is important to emphasize that although the pioneers of the mental testing movement (Terman, Robert Yerkes, and Henry Goddard) were imbued with eugenics, this should not force us to conclude that the study of psychology is racist, biased, or antiegalitarian. It does point out the fine line between psychology and public policy. In the same way there is also a fine line between genetics, biology, and public policy (Provine 1973). The major question is what are the limits of the *implications* of such studies, and not what are the limitations of the studies. A number of rather naive social scientists have condemned doing research in the area of sociobiology because of its apparent disreputable past. This is patently absurd. Given the cultural moment of modern American society, this smacks of the worst sort of authoritarianism. Of course, care must be taken that the implications offered are legitimate and that other, alternate, explanations can see the light of day. When former Attorney General William Saxby stated in 1974 that genes for communism tended to be more frequent in Jewish families, few (we hope) took him seriously.

With this background of the social intent of Darwin, Spencer, Sumner, and the eugenics movement in mind, we will now turn to sociobiology, not as an "enemy of the people," but as a science.

Sociobiology, Ethics, and the Biology of Inequality

> I am saying how things have evolved. I am not saying how
> we humans morally ought to behave.
> Richard Dawkins, *The Selfish Gene*

Introduction

Sociobiology is an attempt to explain the social behavior of organisms in biological terms. In particular, sociobiologists employ neo-Darwinian conceptualizations of natural selection to account for social behavior. In their more ambitious moments, some sociobiologists argue that the subject matter of the social sciences is really biological in character. Although theorists in the field differ among themselves in many important ways, the thrust of their approach is that social phenomena in the animal kingdom, and arguably among humans as well, can be accounted for along evolutionary lines.

A central claim of sociobiology is that organisms act so as to maximize their inclusive fitness. That is, they act to maximize chances of copies of their genes passing into succeeding generations. This hypothesis does not imply that organisms act with the conscious intent of maximizing inclusive fitness or that they would conceptualize their behavior in such terms. The hypothesis purports to explain behavior, not necessarily to imply that organisms conceive of their own actions in sociobiological terms.

Sociobiologists also emphasize the concept of an Evolutionary Stable Strategy (ESS). A strategy for passing on copies of one's genes to future generations is an ESS if it defeats all other strategies, that is, those organisms that act as it dictates will increase their inclusive fitness to a greater extent than will followers of any alternative strategy. Employing primarily the hypothesis of inclusive fitness and the concept of an ESS, sociobiologists attempt to explain the evolution and persistence of various behaviors, including social practices, in terms of natural selection.

The Implications of Sociobiology

Clearly, the most controversial element of sociobiology is the attempt by some sociobiologists to apply their explanatory framework to human behavior. Two related areas of application are of special interest given the subject matter of this particular collection.

The first of these is generated by suggestions to the effect that values, moral codes, ethical principles, and theories are ultimately expressions of biological imperatives. While this thesis is ambiguous, in its more radical versions it suggests that ethics and biology are not distinct in the way that most moralists and moral philosophers would have thought. The second area of application concerns equality or, more accurately, inequality. Some sociobiologists have asserted that certain social inequalities are relatively implastic since they are but manifestations of an underlying genetic core of human nature. These areas are related in that, in each, sociobiology is used to suggest that (to paraphrase Bentham) nature has placed us under sovereign masters, our selfish genes, and it is for them alone to point out what we ought to do as well as determine what we shall do.

In what follows, these applications of sociobiology will be examined. But some preliminary words of caution are in order. First, some sociobiologists are much more cautious in extrapolating to human behavior than are others. Even where less cautious sociobiologists are at issue, difficulties of precise interpretation arise. Accordingly, we will be concerned with what the principles of sociobiology actually imply and not with the writings of any particular theorist. Second, the space available does not allow for exhaustive treatment of all the issues raised. However, enough can be said to suggest how much light sociobiology is likely to shed on the areas under consideration.

Sociobiology, Human Nature, and Ethics

If organisms act to maximize their inclusive fitness, what follows about human nature? What are the implications for our understanding of ethics?[4]

While one might suspect that sociobiology may commit us to a Hobbesian view of organisms struggling in a war of all against all, such a conclusion would be unwarranted. In order to see why, we must distinguish the current notion of Darwinian fitness from the broader modern concept of inclusive fitness. Darwinian fitness is a measure of the individual organism's capacity to produce surviving offspring. Inclusive fitness is a measure of the organism's capacity for having copies of its genes enter future generations. Accordingly, an organism can increase its inclusive fitness, but not its Darwinian fitness, without itself having offspring. It can do this by increasing the reproductive chances of other organisms with which it shares genes.

Given the distinction, sociobiologists leave room in their theoretical approach for what they would call altruistic acts. By an altruistic act, a sociobiologist means something like the following: an act is altruistic if and only if it lowers the Darwinian fitness of the actor while increasing the Darwinian fitness of the recipient of the action. Such behavior is difficult to explain within the context of neo-Darwinian theory, since such theory postulates that organisms act to increase fitness in the traditional, personal reproductive sense. Modern sociobiologists explain such altruistic behavior by arguing that *although altruism decreases the organism's Darwinian fitness, it increases its inclusive fitness!*

Just how is inclusive fitness increased? One attempt at explanation is the theory of kin selection according to which organisms will (or will tend to) engage in altruistic acts when such acts benefit relatives, who presumably share disproportionately the genotype of the agent. For example, since individuals share one-eighth of their genes with their first cousins,[5] the theory of kin selection predicts that they will be (or will tend to be) indifferent between two alternatives, one of which increases a cousin's Darwinian fitness eight times the degree to which the alternative increases their own Darwinian fitness.[6]

In addition to systems of kin selection, sociobiologists also explain some altruistic acts as contributing to inclusive fitness through the operation of systems of reciprocal altruism. In such systems, organisms act altruistically towards others and are themselves the recipients of acts of altruism in return. If the net benefit received in terms of gains in inclusive fitness is on the average greater than the net loss generated by self-sacrifice, one would expect genes favoring reciprocal altruism to spread through the gene pool. Sociobiology predicts, then, that organisms, perhaps including humans, will exhibit altruistic behavior—in the sense explained earlier—to the extent inclusive fitness is thereby increased.

Does it follow that we can totally reject the Hobbesian picture of the state of nature? Not necessarily! For while it is true that reciprocal altruism may pay off, what may pay off even more is a strategy of disguised Darwinian egoism. According to this Thrasymachean alternative, one should appear to be part of the system of reciprocal altruism but, in fact, act only to receive the benefits without sharing in the risks. Since such a strategy would yield all of the benefits and none of the burdens of full participation, one would expect genes favoring disguised egoism to spread rapidly through a gene pool previously dominated by directives favoring reciprocal altruism. Eventually, however, the altruists would find some way of detecting the egoists, in which case disguised egoism would no longer pay. Altruists would only act to favor other altruists, thereby excluding egoists from the benefits of the system. More precisely, genes favoring altruism would have a higher probability of being replicated in the next generation than would genes promoting Darwinian egoism. More elaborate disguises for the egoists might then evolve, be detected and evolve still further until an evolutionary stable system, involving some predictable proportion of egoists to altruists, develops.

Now the logic of the situation described above has been known to theorists at least since the time of Plato's *Republic*. In that dialogue, Thrasymachus challenges Socrates to show that morality pays. On one reading of that challenge, Thrasymachus might acknowledge that all do better under a system of morality—which enjoins other-regarding behavior—than they would do in an egoistic war of all against all. But what Thrasymachus really wants to know is why the strong should not merely appear to be moral, thereby accruing the advantages of the moral behavior of others and becoming objects of admiration besides, while actually acting egoistically whenever self-interest requires it.

Sociobiologists can reply, however, they have at least uncovered the biological basis of the problem and, indeed, of ethics itself. But what can this latter claim mean?

We suggest that it can be understood in at least the three following ways.

C_1 : The principal concepts of ethics, particularly altruism and egoism, are just the concepts sociobiologists deal with. Accordingly, in explaining behavior which falls under those concepts, sociobiologists are thereby explaining the very behavior which is the concern of ethicists.

C_2 : The particular ethical beliefs individuals or groups hold will be (or will tend to be) those that maximize inclusive fitness.

C_3 : The practice of moral evaluation and moral reasoning persists in a given cultural context when and only when it maximizes inclusive fitness of the participants.

C_1 will be called the *conceptual thesis*, while C_2 and C_3 respectively will be referred to as the *strong deterministic thesis* and the *weak deterministic thesis*. "Weak" and "strong" do not refer to the strength of the alleged determinism since both claims may be probabilistic as well as deterministic, but to its scope, i.e., whether what is determined are particular beliefs or a distinctive mode of criticism and evaluation. By "moral evaluation and reasoning," we will mean reasoning from a point of view or set of values and principles that requires us to count the interests of other persons as at least equal to our own.

Our inquiry profitably can begin with examination of the conceptual thesis. According to this thesis, since both ethics and sociobiology are concerned with altruism and egoism, they have the same subject matter. Since they have the same subject matter, sociobiologists are directly accounting for ethical and selfish behavior by tracing its origins to the genetic core of human nature.

In reply to the objection that ethics involves the justification rather than the explanation of moral beliefs, principles, and perspectives, sociobiologists may appeal to the strong deterministic thesis and respond that what is taken to be justification is really that which maximizes inclusive fitness. However, let us leave that claim aside for the moment to concentrate on the thesis that, due to identity of subject matter, sociobiologists have explained the origin of moral behavior in humans by tracing its connection with maximization of inclusive fitness.

The first difficulty with the conceptual thesis is one that applies to the form of sociobiological explanation generally.

That is, the claim that certain behavior is adaptive does not entail that it is genetically induced (Hubbard, 1978). It is at least possible in the case of humans that adaptive behavior is the result of reason or culture, not genes.

Sociobiologists might respond that the way in which reason and culture evolve is itself a result of biology. Once again, let us put aside this claim for later examination and concentrate on the conceptual thesis itself.

According to this thesis, sociobiology and ethics deal principally with the same subject matter: altruistic and egoistic behavior. In fact, ethical theory deals with far more. Questions such as "Should the law interfere with the private actions of consenting adults?" or "Does rule utilitarianism collapse into act utilitarianism?" do not seem to involve the same subject matter as does sociobiology. Moreover, ethicists and sociobiologists ask different sorts of questions, even when egoism and altruism are at stake. The ethicist wants to know if the enterprise of morality has a rational justification while the sociobiologist is interested in explanation of moral behavior.

These complications aside, a far more serious objection warrants attention. For contrary to the conceptual thesis, the subject matter of discussions of altruism and egoism in ethics is different from discussions in sociobiology. That is, "altruism" and "egoism" as used by sociobiologists (henceforth "altruism$_S$" and "egotism$_S$") have entirely different meanings than do "altruism" and "egoism" as used in both ordinary and philosophical moral discourse. Accordingly, arguments which jump from premises about the sociobiological concepts to conclusions about the ordinary ones will be *invalid*.

A few examples should establish this point. Suppose Jones selfishly refuses to share information with you with the intent of advancing his career and retarding yours. In fact, Jones's strategy works and he secures a promotion and raise while you do not. Surely, he has acted egoistically in any ordinary sense of that term. However, a number of members of the opposite sex learn of your plight. Their initial sympathy for you turns to attraction. Their initial attraction towards Jones turns to repulsion. As a result, your Darwinian fitness increases while Jones's decreases. Accordingly, Jones acted altruistically$_S$ since his action results in an increase of your Darwinian fitness at the expense of his own, and egoistically, since he intended to, and indeed did, increase his overall well-being at the expense of yours. Similarly, the very same act can be altruistic in the ordinary moral sense and egoistic$_S$ at the same time. Suppose, to modify the previous example, Jones shares the information with you solely because of concern for your welfare. As a result, you get the promotion and gain far more than he does in terms of the career goals you both value most highly. However, Jones's action becomes known to some and, as a result, others' estimation of his character goes up. He is viewed as more attractive than before, and you, if anything, are viewed as less attractive—a passive beneficiary of the good intentions of another. Hence, his Darwinian fitness increases while yours decreases. Jones's act of information sharing then is altruistic in the ordinary moral sense but egoistic from the sociobiological perspective.

Accordingly, just because sociobiologists and ethicists each speak of "altruism" and "egoism," it does not follow that they are dealing with the same subject matter. There are then good grounds for suspicion of easy or direct transitions from sociobiology to ethics.

However, according to C_2 and C_3, the relationship between sociobiology and ethics is not conceptual but causal. According to C_2, the strong deterministic thesis (SDT), the particular moral beliefs individuals hold will be, or will tend to be, those that maximize inclusive fitness. According to a less radical version of the SDT, culture will indirectly influence what beliefs are held, but the kinds of beliefs emphasized in various cultural contexts will be those that maximize inclusive fitness.

The SDT and the weak deterministic thesis (WDT) differ in the role they assign to culture. The former suggests that the moral perspective of individuals can be explained by direct appeal to maximization of inclusive fitness. The WDT allows that individual moral outlooks are influenced by cultural factors but add that if we want to explain why culture has developed in the indicated way, the rock-bottom explanation is in terms of maximization of inclusive fitness.

How is the SDT to be evaluated? Consider Wilson's remark

in *Sociobiology* that ethicists may misidentify as intuitive perceptions of basic moral principles what in reality are "the emotive centers of their own hypothalamic-limbic center" (Wilson, 1975). This hypothalamic-limbic center is itself, of course, the product of evolution.

What is the import of Wilson's remark? Although what he means by it is far from clear, one interpretation is that what we take to be critical examination of a moral problem is really nothing more than a biologically conditioned response. On this reading, evolutionary explanations undermine the rationality of ethical inquiry. What we take to be rational examination of a moral issue is in reality nothing but the clicking out of our genetic program.

Unfortunately for Wilson, if this is his point, he would be foolish to press it. For if rational inquiry is preprogrammed biologically to come to certain conclusions regardless of the actual strength of the evidence, what happens to his own theory Surely he would claim that sociobiology is supported by the evidence. But if sociobiology shows that we really do not consider the evidence but only appear to do so in a biologically determined way, then we do not really consider the evidence for sociobiology itself and have no rational grounds for acceptin it, or anything else for that matter.[7]

On the other hand, if we read Wilson as more plausibly asserting that although we are capable of being guided by reason, the values we hold in fact are the result of the evolutionary process, there is no special cause for concern. While such a thesis may be of scientific interest, it no more threatens the distinctively ethical enterprise than does the view that the values we hold are the result of our being brought up in a certain culture or in a certain socioeconomic class. In either case, rational examination and criticism of current values are both possible and causally efficacious. If they are impossible, so is the rational justification of sociobiology.

Accordingly, we conclude that on the first interpretation, the SDT undermines the case for sociobiology itself. On the other hand, the second interpretation not only does not deny that rational inquiry can be causally efficacious; it provides still another reason for engaging in it, i.e., to discover unjustified but genetically induced moral judgments.

The *weak deterministic thesis* (WDT) does not deny either that we can or often are moved to accept particular moral judgments, principles, and theories as a result of reasoned inquiry. Rather, its distinctive claim is that the rock-bottom explanation of why we think in moral terms at all is ultimately biological. Proponents of the WDT can concede that "altruism" has different meanings in ethics and sociobiology. But they go on to assert that the ultimate cause of our employment of a distinctively moral mode of evaluation is sociobiological.

Once again, we have a claim which is scientifically interesting but from which no extreme social implications follow. Proponents of the SDT acknowledge and may insist that moral inquiry and moral reasoning can influence conduct. Their point is not that such moral reasoning is biologically directed to incline in particular directions, but rather that the proclivity to engage in and be influenced by it has a biological origin. Such a view may or may not be true but does nothing to establish that our moral life is simply a superstructure, the nature of which is in reality dictated by genetically grounded biological imperatives. Similarly, if it were true that our tendency to engage in scientific inquiry was genetically induced, this would do nothing to collapse the distinction between acceptable and unacceptable scientific theories.

Dawkins, in his book *The Selfish Gene*, suggests than an organism relatively free of a direct genetic program for behavior will have enormous adaptive advantages over organisms genetically "wired" to react to a limited range of environmental problems in specific ways. He postulates that humans may be organisms whose intelligence autonomously can set priorities for action.[8] If this view is rejected, there is no reason to believe that sociobiology itself rests on a rational foundation.

Sociobiology and Inequality

Sociobiology has perhaps proved most controversial when applied to questions of equality and inequality among human beings. Wilson in particular has suggested that certain sorts of inequalities among humans, including sex inequalities, may have a genetic origin (Wilson, 1978). In what follows, we will focus not on the views of a particular sociobiologist such as

Wilson, but on what one is entitled to infer from the principles of sociobiology.

Drawing on the presentation in the last section, we will suggest that discussions of human inequality based on sociobiology are likely to make two serious kinds of mistakes. Relying on the dubious conceptual thesis, sociobiologists may illegitimately jump from premises containing only sociobiological concepts to conclusions containing radically different concepts, only superficially resembling those in the premises. Relying on the equally dubious SDT, sociobiologists may maintain that because a certain practice arises from deep, genetically rooted impulses, it is relatively impervious to social criticism and social change.

While this section will focus on the two points sketched above, an additional difficulty must be kept in mind. The claim that complex human behaviors have a genetic origin is itself exceedingly controversial. As we have seen, it is all too tempting to infer the conclusion that a particular behavior is genetically induced from premises establishing its adaptiveness. Such reasoning ignores the existence of alternative forms of explanation and in effect dismisses them without a fair hearing.

Be that as it may, the discussion in the last section suggests that even if genetically induced behavioral tendencies exist, sweeping social implications need not follow. Consider, for example, claims about sex differences found in the sociobiological literature. Sociobiologists have suggested that if they are to maximize inclusive fitness, male and female organisms should adopt significantly different strategies towards reproduction.[9] Specifically, males should try to impregnate as many females as possible without committing themselves to any particular female since even if a relatively low percentage of offspring survive, the total number will be high. If females are to get copies of their genes into the next generation, they have no choice but to care for the offspring of the unfaithful male. The best strategy for a female, then, is to invest a considerable amount of time and energy in making sure a high percentage of the relatively few offspring she can bear survive. In addition, females profit if they can encourage males to help with the task of caring for offspring. Accordingly, female traits which encourage commitment on the part of males would be favored by natural selection. (For discussion, see Dawkins, 1976, chapter 9). What implications does such a view have for sex inequality among human beings? Some sociobiologists are reluctant to extrapolate to humans while others show no such hesitation. In the face of such disagreement among sociobiologists themselves, it will be useful to consider difficulties that must be faced by those who would extend their conclusions to the human arena. Suppose it were claimed, for example, that human females are genetically inclined to adopt certain social roles, particularly those that emphasize nurturing of a family, or at least to avoid certain other kinds of social roles, particularly those that involve assumption of high risks. Is there any warrant for such a claim?

One objection that proponents of such a view must face is that their position seems to rest on (or on something very much like) what we have called the conceptual thesis. For the claim in question involves a questionable jump from a sociobiological premise containing sociobiological concepts to a societal conclusion containing concepts that are not sociobiological in character. For what is assumed is that a biologically conservative strategy for maximizing fitness will also be a socially conservative one. But contrary to that supposition, biological and social conservatism are radically different. For example, the best strategy for maximizing female fitness in contemporary Western societies may be to excel at some career and become socially, politically, and economically independent. Arguably, such an independent woman is best able to get copies of her genes into the next generation—either by contributing to a system of reciprocal altruism or by providing better care for the offspring she chooses to have—at least in the highly technological, postindustrial societies of the West. Accordingly, one cannot jump directly from premises about which strategies maximize inclusive fitness to conclusions about the biological functions served by existing institutions. Depending upon the social context, alteration of the institution may be the best strategy some individuals can adopt if they wish to maximize their own fitness.

Even more to the point, humans generally do not act with

the conscious intent of increasing inclusive fitness. As we suggested in our discussion of the SDT, moral criticism and evaluation can be among the factors which influence human behavior Indeed, moral criticism, by altering the environment, can determine which genes will be replicated in future generations.

Suppose, for example, that it is conceded, if only for the sake of argument, that a disproportionate number of women feel uncomfortable in traditional male roles and that this discomfort is due predominantly to their genetic heritage. But then, due to moral criticism of traditional sex roles, their repudiation by women increasingly is viewed as desirable. As a result, women who do not fit traditional models are seen as more and more attractive by males. But these women presumably are those who are least coded by their genes for traditional behavior. To the extent such women gain in Darwinian fitness, they will pass on copies of their "liberated" genes more frequently than before. Accordingly, copies of their genes increase in the overall gene pool and traditional sex inequalities tend to break down even further.

This suggests that rather than ethics being an epiphenomenon of natural selection, it can determine the course selection takes. The temptation expressed in the SDT is to be resisted. Genetic explanations do not imply that the behavior explained is implastic. Rather, by altering environments, human agents can determine what kinds of effects our selfish genes may promote.

We conclude that even if complex human behaviors are genetically induced, they need not be impervious to social change or moral criticism. Sociobiology will not solve our ethical problems for us. Still less will it show that certain problems must be forever with us, as part of our genetic heritage. Rather than revealing that selfish genes are our sovereign masters, a proper interpretation of sociobiology suggests that it is we, ourselves, who can and should determine the rightful boundaries of their domain.

Notes

1. This notion of history also complicates any lawlike physical or chemical findings in biology. "One could at the same time assume, as Bohr

has suggested, that our knowledge of a cell being alive may be complementary to the complete knowledge of its molecular structure. Since a complete knowledge of this structure could possibly be achieved only by operations that destroy the life of the cell, it is logically possible that life precludes the complete determination of its underlying physico-chemical structure" (Heisenberg, 1962:104-5).

2. This should not suggest that individuals like Agassiz were any more egalitarian in their focus. In fact, Agassiz wrote several treatises warning against teaching Negroes too much for fear that the pressure would crack their crania and kill them.

3. "Methodological superiority, like any other, is relative to the end. If the end is control and prediction, then faith through critical selection of evidence is superior. If, on the other hand, the end is the development of aesthetic discrimination, or the intensification of a sense of holiness, or the mixture of apocalyptic expectation, then methodologically the technique and ruthlessness of blind faith may well be the superior means. Dedication to one end rather than another, and accordingly to one means rather than another, is a problem of moral choice" (Buchler, 1951:166).

4. Much of the material on pp. 52-62 first appeared in Simon and Zegura (1979). We are grateful to *Social Research* for permission to use this material here.

5. Over and above the common stock of genes shared by all organisms of the same species.

6. The concept of kin selection and Hamilton's rule (benefit to recipient of altruistic act over cost to performer of the act must be greater than $1/r$, where r is the coefficient of relationship) are due to the work of Haldane (1932, 1955) and Hamilton (1964).

7. While it is logically possible that reasoning in ethics is genetically directed but reasoning in science is not, it surely is arbitrary and ad hoc to postulate such a distinction simply in order to save Wilson from criticism.

8. The point here is not that reasoning is causally undermined but rather that logical considerations may act as causes!

9. At least in species where the female is capable of bearing only relatively few offspring.

4

The Issue of Inequality: Herbert Spencer and the Politics of the New Conservatives

STUART C. GILMAN

A father said to his double-seeing son, "Son, you see two
instead of one."
"How can that be?" the boy replied. "If I were, there would
seem to be four moons up there in place of two."
Idries Shah, *Caravan of Dreams*

Men make their own history, but not of their own free will. . . .
The tradition of the dead generations weighs like a nightmare
on the minds of the living.
Karl Marx, *The Eighteenth
Brumaire of Louis Bonaparte*

This is an essay on "seeing" equality. In this sense there will
be no attempt to extrapolate equality in its social relations (Rees
1971) or its anthropological origins (Rousseau, 1964). Rather,
the thrust of this essay will be to understand the development
of a specific view of equality in contemporary social science,
tracing its roots into nineteenth-century sociology, and suggesting some contemporary implications. The subject matter of
this essay is the philosophical underpinnings of the "new conservatives" (Coser and Howe, 1977) or those social scientists who
support what can be broadly classified as the "social inequality"
thesis. The focus will be on these individuals' understanding of
human being, that is, its ontological status and its impact on
their empirical findings and subsequent sociological arguments.

Before engaging our subject, it is important to emphasize that the process and epistemology unearthed in the following discussion is ultimately political. That is, in a broad sense this is a study in the politics of the sociology of knowledge. The movement herein described has created the basis for political argument and political action. It has led to a significant undercurrent in the social sciences which supports a rather stultifying eugenics, in turn promoting a public policy in which benign neglect is often the ascriptive metaphor. It has succeeded in muting the egalitarian thrust of the 1960s by substituting an ideological given: inferiority. This has subsequently been cited as sociological truth in many of the political science texts which deal with the subject.[1] The weapon used in capturing this ideological moment was statistical analysis—more appropriately data analysis—which was the same weapon which justified equality a decade before. These tools are not neutral but rather plastic and have been used to present fantasy as fact while shrouding the notion of equality in a language both foreign and capable of arbitrary interpretation. This is a political act, albeit conscious or unconscious, and not mere sociological or psychological analysis.

The Cast(e) of Characters

The new conservatives present an impressive issue of names and reputations: Nathan Glazer, Daniel Moynihan, James Coleman, Seymour Martin Lipset, Richard Herrnstein, Irving Kristol, and Thomas Sowell, to name just a few. How can one lump all of these individuals with their disparate specialties, disciplines, and even interests into one category? Joseph Epstein (Coser and Howe, 1977:9-28) suggests that they were victims of a conservative plague which eroded the health of their previous commitment to "New Deal" liberalism. They got "cold feet in the 1960s." This appears a bit simplistic and can only appeal to a lemminglike mass insanity for explanation.

Certainly, a paradigm shift could explain it. Unfortunately, Professor Kuhn's notion of paradigm and scientific revolution breaks down because the case of the new conservatives is not a question of advancement, but a reevaluation of the data itself.

That is, it is not the arising of new anomalies, but rather the re-creation of the old ones. Additionally, the notion of paradigm has come upon hard times in the houses of the historians and philosophers of sciences. (Perhaps I intuitively knew of its loss of utility several years ago when my fellow graduate student began inquiring, "Brother, can you paradigm?") For our immediate concern here I would like to suggest the use of Stephen Pepper's notion of "root-metaphor." A root-metaphor is the common starting point from which individuals begin to view the world and its social artifacts.

> A man desiring to understand the world looks about
> for a clue to its comprehension. He pitches upon
> some area of commonsense fact and tries (to see)
> if he cannot understand other areas in terms of this
> one. This original area becomes then his basic ana-
> logy or root-metaphor. He describes as best he can
> the characteristics of this area, or, if you will, dis-
> criminates its structure. A list of its structural char-
> acteristics becomes his basic concepts of explanation
> and description. We call them a set of categories. In
> terms of these categories he proceeds to study all
> other areas of fact whether uncriticized or previously
> criticized. He undertakes to interpret all facts in terms
> of these categories. As a result of the impact of these
> other facts upon his categories, he may qualify and
> readjust the categories, so that a set of categories
> commonly changes and develops. (Pepper, 1966:91)

Although the categories will expand to take advantage of new facts, the root-metaphor which derives out of common sense maintains itself, becoming more or less useful with expansion of danda, Pepper's term for interpreted facts.

This common starting point is the notion of equality itself. As Robert Nisbet points out, equality always comes into conflict with freedom, and this is the essential political quandary which social scientists such as Christopher Jencks and John Rawls fail to resolve (Nisbet, 1974). Nisbet gives one of the best "new conservative" arguments for delimiting equality and making choices about the goodness of equality as it exists in modern society. The point is that the situation of equality presently existing, coined meritocracy, is the ultimate end of modern technological society.[2] Certainly, adjustments will be made within their scheme of things, but essentially equality is defined as an attribute of what currently holds force in American society. Additionally, it is a circumstance with normative justification which is lacking in other conceptions of equality—most notably Marxism (Bell, 1974). Thus, equality, which is both a root-metaphor and the ontological basis of politics for the new conservatives, will be shown to determine the categories of sociological analysis for this new sociological movement.

The notion of root-metaphor allows us to develop the ontological basis of the new conservatives, while presenting an analytical framework through which one can begin to understand why they must begin from the initial premise of "inequality." This should not be taken to suggest that all of these perspectives are identical, but, at the risk of redundancy, their analysis has the same view of man. What follows, then, is a characterization of this view. Within this perspective, there will be no attempt to criticize the new conservative analysis as to "factual" interpretation. This is a project which has already been undertaken by others.[3] Rather, our view will be delimited by the notion of political "being" and its political ground (Heidegger, 1961). That is, how is man's political existence viewed in the analytical "drawings" of the new conservatives?

The Ontology of Political Inequality

For the meritocrat, inequality does not exist without purpose in society. In this sense there is no difference between his position and that of the most ardent egalitarian. The difference between them is created in the cause of inequality.

> One sees this (the populist argument—qua egalitarian) today in the derogation of the IQ and the denunciation of theories espousing a genetic basis of intelligence; in the demand for "open admission" to universities; in the pressure for increased numbers of

blacks, women, and specific minority groups such as
Puerto Ricans and Chicanos on the faculties of
universities, by quotas if necessary; and in the attack
on "credentials" and even schooling itself as the
determinant of a man's position in society. (Bell,
1972:31)

Inequality is thus inherent in the nature of man which only the
most ardent utopians would deny. Taking a rather sour note
from Rousseau, they point out that man in modern society does
have unequal capacities. If these are ignored, then there is no
longer any basis for reward; hence, progress and development
in society will be stultified. Man's nature is evolving toward a
more and more perfect state in which those of the greatest
competence will get the most reward. In fact, Bell grounds his
entire argument on Michael Young's fable, *The Rise of Meritocracy*, which moralizes that "Civilization does not depend upon
the stolid mass, *the homme moyen sensuel*, but upon the creative minority, the innovator who with one stroke can save the
labour of 10,000, the brilliant few who cannot look without
wonder. . . . Progress is their triumph; the modern world their
monument" (Bell, 1972:fn. 2). Thus, meritocracy is not only
a natural evolving of society, but also its goal, its end, its consum
mation. In other words, the first element of the new conservatives' root-metaphor is the natural state of meritocratic society.
Man naturally exists in this highly competitive framework.

Richard Herrnstein lays another of the ontological roots for
meritocracy by firmly planting the success of humans in the
genetic development of IQ. In his now classic article, Herrnstein
(1971a:51) argues that intelligence measurement is not only
psychology's most telling accomplishment to date, but also has
been determined to be so objectively sound that it "cannot be
repudiated on technical grounds alone."[4] He goes on to argue
that there is a significant correlation between class and IQ in
the United States and concludes that "it is probably no mere
coincidence that those values (meritocracy) often put the
bright people in the prestigious jobs. By doing so, society
expresses its recognition, however imprecise, of the importance
and scarcity of intellectual ability" (Herrnstein, 1971:51). Rely-

ing heavily on the arguments put forth in Arthur Jensen's (1969) controversial work, Herrnstein concludes that the body of psychological studies support the inheritance of intelligence.

The most useful data for Jensen and Herrnstein come from the now infamous twin studies "which set genetic similarity high and environmental similarity low."[5] Herrnstein points out that this genetic similarity inevitably leads to the question of race. Although he avoids the issue by noting that sufficient research has yet to be done to establish such a conclusion, he does infer that Jensen quotient of heritability of .8 (80 percent of all intelligence is inherited) is highly suggestive "of a genetic component in the black-white difference."

The point of Herrnstein's article, and his work in this area generally, is that nature is responsible for intelligence. Both he and others of the new conservative persuasion use this as a basis for arguing that compensatory education will not work. This is again interpreted to suggest that all but the most basic juridical compensation (e.g., the doing away with segregation laws) will have no impact. The concluding line of this argument would suggest that equality itself was a myth, however well intentioned (Nisbet, 1974), and we must learn to accept this description of the nature of man. Herrnstein argues that this is not an attempt to create Galton's vision of eugenics, for "no sensible person would want to entrust state-run human breeding to those who control today's states." His is a subtler form of eugenics by which criteria can be developed to determine intelligence (Herrnstein, 1971a:58).

It is important to emphasize that the ontological basis of man is fixed and ontic in the sense which the new conservatives have portrayed it. It is not that man will continue to develop, but rather the way certain *men* continue to develop. The point is that for the majority of society, evolution is predetermined although not necessarily static. Herrnstein relies heavily upon Louis Terman, who was the first to view the relationship between IQ and social class in his five-volume work *Genetic Studies of Genius* (1925). He suggests that Terman was the intellectual grandfather of Jensen, and in grounding the tradition both Jensen and Herrnstein espouse he was the first to "objectively" determine the importance of social class as an ascriptive variable (i.e., independent) of IQ. But, Terman and his associates had specific purposes in mind for the use of his studies. Terman conducted a study of Italian immigrants and their children in which he concluded that Italians had an average IQ of 84. They were just one of man's genetically inferior "races" and they should be segregated into special classes:

> They cannot master abstractions, but they can often be made efficient workers. . . . There is no possibility at present of convincing society that they should not be allowed to reproduce, although from a eugenic point of view they constitute a grave problem because of their unusually prolific breeding.[6]

The point Terman makes is that there is a necessary political logic once one has determined genetic inferiority. This is not to suggest that the meritocrats are stoking the crematoriums. Quite the contrary, they are laying the foundation by which society can accept itself in its present form. The meritocratic society propounds that those who should get ahead do, the point being that it is a crucial error to assume that this is a reactionary position. In fact, it is truly "liberal." The implications of this will be explored in more detail below.

Our first thesis attributed to the "new conservatives" is that meritocracy is both the moral and the natural "state" of man. I use "state" in its Greek sense as the original and psychologically fruitful way man can be (D'Entreves, 1967). Second, man is, by his very nature, intellectually predestined to succeed or fail within the context of this meritocratic society. That is, within a society which emphasizes human development and this natural evolution of the best men into the most prestigious positions, they are rewarded rightfully because of their abilities. This brings us to the last ontological premise. Since meritocracy is "demonstrated" to be natural and moral vis-à-vis the constitution of man, and since man's innate intelligence also naturally tends for superior people, albeit due to class or race, to succeed, what the meritocrats must now demonstrate is that modern American—or Western—society replicates this status, or at least comes close. That is, the present environment provides the political and social support for the established ontological basis of man.

In fact, most of the meritocratic literature is specifically devoted to demonstrating this condition.[7] Yet the argument cannot be said to be naive because it intends to show this as the "best of all possible worlds." Rather, it is an argument which attempts to demonstrate that the contemporary American system of government is good enough to be defended against the abrasive attacks which the radicals leveled in the 1960s. That is, in the trade-off of ideas, the meritocratic ideal, née freedom, even with its limitations on equality, is better than the egalitarianism proffered by the left. It is the trade-off of freedom for equality which is bemoaned most. "The outcomes of competition between individuals," writes Daniel Bell,

> are disparate degrees of status, income, and authority. These disparate outcomes have been justified on the ground that they have been freely gained and have been earned by effort. This is the basis of the idea of "just meritocracy," and historically, the striving for liberty *and* equality. . . . The effort to reduce disparates of outcomes means that the liberty of *some* is qualified or sacrificed to make *others* the more equal to them. (Bell, 1974:57)

It is not that Bell denies any adjustments in favor of equality, but rather that these must be minimal because they inevitably lead to "administrative determinism."

The crucial element within this argument is the notion that equality necessarily has a negative impact upon freedom as a human condition. This is not equality at all but intellectual caveat. For this reason concepts such as institutional racism, affirmative action, and the like are dangerous because they tend to make any distinctions unjust or illegal (Glazer, 1975). Distinction is therefore set up as an ontological notion which should predetermine public policy not only because of its roots in human nature, but also because of majority will:

On the evidence of polls and surveys there is little respect for the kind of equalitarianism that matter most to intellectuals; equalitarianism that would by design sweep away the built-in inequalities of family, of inheritance, of luck, and of individual ability and aptitude. To most people, legitimate equality is epitomized by equality of *opportunity* for the great diversity of tastes, talents, strengths, and aspirations to be found in a population. But to a rising number of intellectuals this is the worst kind of *inequality*, for it produces, it is said, a meritocracy, which is in its own way as evil as any of the historical forms of natural privilege.

Majority will, the historical foundation of democracy, cannot, then, be counted on to inaugurate the regime of equality that is desired by intellectuals. . . . The politics of virtue from Plato to Rousseau has rarely coincided in the past with anything easily describable as democracy.[8]

Thus, man not only naturally but also democratically exists in the meritocratic environment. The democratic element is actually innate in his nature.

What is most "terrible" about the egalitarian thrust is that it forces an inquiry into the fundamental values of meritocracy, which tends to undermine society. "There is a widespread questioning of the legitimacy of institutions, especially on the part of the young who would normally move into elite positions— this is the loss of *civitas*" (Bell, 1974:45). The loss of civitas, or what Bell seeks to characterize as civitas, is dreadful because it questions the legitimacy of society. This is wrong by definition because society is good in its very essence, that essence being its foundation in nature. For if man in his social situation is naturally selected, then an individual rejection of that position endangers that essential evolutionary trend of society. This means regress in the face of attainable progress. This is a society characterized as insane. Again picking up Bell's argument: "The foundation of any liberal society is the willingness of all groups to compromise private ends for the public interest. The loss of civitas means either that interests become so polarized, and passions so inflamed, that terrorism and group fighting ensue, and political *anomia* prevails, or that every public exchange becomes a cynical deal in which the most powerful segments benefit at the expense of the weak" (Bell, 1974:46). Thus, individuals now have an obligation—*fidès*—to support this natural status of man.

This is not, as advertised (Bell, 1972:40), an argument for individualism, but rather a radical recasting of what being an individual entails. This is important not in critique, but explanation. For the meritocrats, man has a special cast and nature which becomes closely attached to his social, political, and economic situation. It is not his individuality which is protected, but rather his status. That is, the meritocrat insists that there is something inherent in nature—IQ, genetics, or even luck— which allows for the best development of society. Therefore, gains, albeit social or economic, are not merely rewards, but evidence of natural progress. That is, the full development of society will allow individuals to evolve to the best status possible—at least from the point of view of meritocratic society. Obligation then is not mere self-interest, but rather a normative compulsion to progress.

An attempt has been made to give a brief sketch of the three ontological arguments upon which the meritocratic position stands: (1) its moral status in relation to other forms of sociopolitical organization; (2) intellect does and should determine success; and (3) modern American society replicates the two previous conditions. The criticisms of such a position are evident, if not always victorious. What strikes this author most about this exploration is not its stringent tones, but rather its haunting melody. The ontological argument found here is a variation on a very familiar theme. Its best known form is found in the writings of Herbert Spencer. This is not to say that in their similarity the meritocrat's argument should be dismissed. This is an illegitimate tactic, as Michel Foucault has pointed out:

> In what sense and in accordance with what criteria
> can one affirm: "this has been said"; "the same thing
> can already be found in this or that text," etc.? What

is identity, partial or total, in the order of discourse?
The fact that two enunciations are exactly identical,
that they are made up of the same words used with
the same meaning, does not, as we know, mean that
they are absolutely identical. (Foucault, 1972:143)

One legitimate method is to look at the implications of such
theories and the perspective which their root-metaphors provide.
It is with this in mind that we must now turn to Spencer.

Spencer: The Libertarian Root-Metaphor

There is today a resuscitation of interest in the sociology of
Herbert Spencer.

Perrin (1976:1339)

So begins Robert Perrin's recent article on Spencer. Perhaps it
is mere coincidence that his first note expresses appreciation
to Robert A. Nisbet for his encouragement in his project. Nonetheless, it is true that Spencer enjoys a new popularity and has
developed adherents across the social sciences. The question is
what does Spencer "bring" to his studies, i.e., what is his root-
metaphor?

Most recent writing about Spencer is concerned with his
"functionalism" or "environmental determinism."[9] It was, of
course, Spencer and not Darwin who coined the term *evolution*
(Carneiro, 1973:77-95). However, its interpretation in terms
of "modernisms" like funtionalism is cute but not very productive. Evolution logically, and teleologically, precedes functionalism and its spin-offs. The macro-theories in the social
sciences might indeed prove worthwhile some day, but it seems
garrulous to attempt to grasp Spencer's intent ex post facto.
His meaning must be derived in how he defines evolution and
the root-metaphor in which he anchors it.

One of the greatest curiosities about Spencer—and something
which separates him from the vast majority of social theorists—
is that everything is interpreted in terms of evolution. One
might argue that a number of *social* theorists have this myopic
and singular vision, but that is precisely the point. For Spencer
it was not merely social events which are explained in their

evolutionary intent, but *everything*. As Perrin points out, Spencer applied evolution "from (the) inorganic to (the) superorganic" (Perrin, 1976:1356); that is, from geology, to biology, to spiritual beliefs, to mutilation, to politics, to . . . even sociology. For Spencer, evolution was the demonstrable epistemology for the world. Whether one accepts Perrin's thesis that Spencer had four theories of evolution,[10] or merely one, the point is that the idea of evolution allowed Spencer to dismiss the prominent perspectives of his time, natural rights and utilitarianism, for the sinecure of the progressing universe.

For Spencer the crucial element in ensuring evolution, and its natural moral superiority, is his theory of freedom "that every man may claim the fullest liberty to exercise his faculties compatible with the possession of like liberty by every other man" (Spencer, 1896b:36). He dismisses the idea of limiting liberty to the hurt or pain of the other because it naturally leads to a utilitarian logic and a socialist outcome. His is a direct revision of the notion of equality *in* the liberal tradition. He provides a great number of examples illustrating the fallacy of distributing rewards, rights, and prerogatives from a framework of equality. The only equality than can be hoped for is equality of liberty, translated in the contemporary debate as equality of opportunity. Thus, equality as foreseen by socialism and communism is "impracticable" and "immoral":

> If an equal portion of the earth's produce is awarded to every man, irrespective of the amount of quality of the labour he has contributed towards the obtainment of that produce, a breach of equity is committed. Our first principle (his theory of freedom) requires, not that all shall have like shares of things which minister to the gratification of the faculties, but that all shall have like freedoms to pursue those things—shall have like scope. It is one thing to give to each an opportunity of acquiring the objects he desires; it is another, and quite a different thing, to give the objects themselves, no matter whether due endeavour has or has not been made to them. Nay more, it necessitates an absolute violation of principle of equal freedom. (Spencer, 1896b:65-66)

The keystone to all of Spencer's social analysis is his vision of equality. If evolution were to be the ideographic paradigm, then equality, as it was being interpreted in the social philosophy and policy of his time, would have to be a pernicious idea (Polanyi, 1975:145-46).

Again the argument is returned to how this vision of equality projected itself for Spencer and how evolution had to be tied to the root-metaphor of equality in order for his sociology to "work." The first premise, the natural state of equality, is expressed in his theory of freedom. This is a theory which entails a series of causal assumptions about man and, ultimately, the study of man. Spencer argues that there are a multiplicity of forces which create individuals in society, and he spends several pages enumerating them (Spencer, 1896c:16-20). The point for him is to wade through the various causes, as would a physician, to find the nature of the animal and, hopefully, the cause of the disease. In doing so, Spencer weaves his cloth with the fundamental assertion that social equality was inequity, and freedom was the only true form of equality. The result of attempting social equality was the destruction of freedom in society. "Wherever requirements which have their roots in the order of Nature come to be enforced by an extrinsic authority, obedience to that extrinsic authority takes the place of obedience to the natural requirements." (Spencer, 1, 1896a:542) Thus, state intervention in human activity destroys man's ability to maintain his freedom because he must rely on something outside himself in order to act socially or politically.

Man exists in this competitive environment naturally. He must have natural opposition or progress will not occur. "It seems, from one point of view," writes Spencer,

> unjust that the inferior should be left to suffer the
> evils of their inferiority, for which they are not
> responsible. Nature, which everywhere carries on
> the struggle for life with unqualified severity, so as
> even to prompt the generalization—"the law of
> murder is the law of growth," cares not for the
> claims of the weaker, even to the extent of securing
> them fair-play; and if it be admitted that this severity

of Nature may, among associated men, rightly be
mitigated by artificially securing fair-play to the
inferior, why ought it not to be further mitigated
by saving them from all those evils of inferiority
which may be artificially removed? (Spencer, II,
1896a:273-74)

Of course, artificially removing all obstacles would be pure
equality—illogical and impractical. Most important of all, it
would violate man's very nature and rob him of his chance for
real freedom.

The natural status of man is a crucial element in his root-
metaphor of equality. For not only is he morally and socially
superior through his success in society, but certain men are also
superior to any of those who have not succeeded, e.g., women.
Anthropologically, women have always been inferior for "a
Fijian . . . he might kill and eat his wife if he pleased," and the
"wilder" Australians "sacrificed their old women for food"
(Spencer, II, 1896a:274). (*Post hoc, ergo propter hoc.*) In the
case of women, the very nature and ability of man makes them
superior over women and only in a state of permanent peace
could women even entertain equal political rights.[11] This natural superiority of certain men is also expressed over racial and
class distinctions. Although Spencer deplores class egoism, he
does admit that it has a biological basis, even if it is not in
terribly good taste. Race is also one of the distinguishing characteristics of nature.

While man is mutable, evolution is immutable. Therefore,
man can change, but only in accordance with the paths outlined
for him by nature. By differentiating various statuses of individuals, ordering them into phyla, and describing the resulting
hierarchies, Spencer presents us with an analytical perspective
resulting directly from his root-metaphor: the natural inequality
of man. Given this, he must also demonstrate that there is some
sort of justice in nature.

To describe this justice Spencer falls back on two interrelated
arguments. First, that this is a moral state of man and that tam-
pering with it will only endanger the precious freedoms now
available. Using his organic analogy, Spencer makes arguments

defending this position including man's rights of success, liberty, and, ultimately, the notion of property:

> If, as M. Proudhon asserts, "all property is robbery"— if no one can equitably become the exclusive posses- sor of any article, or as we say, obtain a right to it— then, among other consequences, it follows that a man can have no right to the things he consumes for food. And if these are not his before eating then, how can they become his at all? As Locke asks, "when do they begin to be his? when he digests? or when he eats? or when he boils? or when he brings them home?" If no previous acts can make them his property, neither can any process of assimilation do it: not even absorption of them into the tissues. Wherefore, pursuing the idea, we arrive at the curious conclusion, that as the whole of his bones, muscles, skin, etc., have been thus built up from nutriment not belonging to him, a man has no property in his own flesh and blood—has no more claim to his own limbs than he has to the limbs of another; and has as good a right to his neighbor's body as his own! (Spencer, 1896b:66-67)

Property becomes the paradigmatic symbol of man's success and of the natural morality inherent in a capitalism which guarantees equal liberties. It would be in error to assume that this is an argument for a natural elite who would hand down from generation to generation their nobility. Rather, it is an argument for seeing the mechanisms of a free society as a tool for creating individuality and a sepulcher of morality. The hierarchy which Spencer describes is not necessarily one of inheritance, that is, title, but rather one bestowed by nature in the very forces of the community.

Man exists as an individual in this structure. His second point attaches individuality to intellect, thereby creating intelligence as the ultimate distinction among men. He "is furthest removed from the inorganic world in which there is least individuality. Again, his intelligence and adaptability commonly enable him to maintain life to an old age . . . that is, to fill out the limits of his individuality to the full. Again, he is self-conscious; that is, he recognizes his own individuality" (Spencer, 1896b:258). In contradistinction to the meritocrats, Spencer does not believe that there is any way to distribute liberties in order to ensure the supremacy of the best. Rather, the natural processes of the free society must be relied upon. Like the meritocrats he believes that society, as it exists, distributes freedoms as best as can be expected. However, the most singular similarity is in the view that it is not society which is the driving force behind history. Rather, it is the exceptional individual who will not consent to be reduced to the level of their inferior contemporaries (Lakoff, 1964:150-51).

For Spencer, intelligence induces moral conduct, and average intelligence—what he characterizes as the goal of the equalitarians—is incapable of guiding even simple conduct. "The unthinking ineptitude with which even the routine of life is carried on by the mass of men, shows clearly that they have nothing like the insight required for self-guidance in the absence of an authoritative code of conduct" (Spencer, 1896c:303). Although there is no way to determine intelligence or merit, it is important to emphasize it as a fundamental criteria for governing society.[12] Since intelligence is both ontogenic and phylogenic, Spencer also perceives fundamental cultural and intellectual differences among peoples. Subjection "to different modes of life, produces in course of ages permanent bodily and mental differences: The Hindu and the Englishman, the Greek and the Dutchman, have acquired undeniable contrasts of nature, physical and psychical, which can be ascribed to nothing but the continuous effects of circumstances, material, moral, and social, on the activities and therefore on the constitution" (Spencer, 1896c:338). Intelligence is the element which the mechanisms of society naturally use to distinguish the success of men. Since there is no governmental way to ensure these differences, man must rely on the justice of nature and his own moral instincts.

Spencer's greatest concern in his writings, as with the meritocrats, is the criticism of liberalism for encouraging governmental expansion at the cost of retreating from a commitment to individuality and freedom. This gives us the third element of his root-metaphor. Spencer reacted indignantly to John Stuart Mill, as well as to liberal politicians of his day, because he gave up the principles of liberalism, notably individualism, for the dreamworld of the welfare state (Spencer, 1896b:281-311). "How is it that Liberalism, getting more and more into power, has grown more and more coercive in its legislation? How is it that, either directly through its own majorities or its opponents, Liberalism has to an increasing extent adopted the policy of dictating the actions of citizens, and, by consequence, diminishing the range throughout which their actions remain free?" (Spencer, 1896b:285). For Spencer the reason was clear. The liberals had lost the ability to accept the morality in the status of the individual. The increasing misery of the workers was an inevitable outcome of true evolutionary development.

There can be sympathy for the individuals left behind by social evolution, but any attempt to allow government to interfere in this natural condition would lead to disaster.

> Given an average defect of nature among the units of
> a society, and no skilful [sic] manipulation of them
> will prevent that defect from producing bad results.
> It is possible to change the form of these bad results;
> it is possible to change the places at which they are
> manifested; but it is not possible to get rid of them.
> The belief that faulty character can so organize it-
> self socially, as to get out of itself a conduct which
> is not proportionately faulty, is an utterly baseless
> belief. . . . Where the evil does not, . . . reappear in an-
> other place or form, it is necessarily felt in the shape
> of a diffused privation. For suppose that by some of-
> ficial instrumentality you actually suppress an evil, in-
> stead of thrusting it from one spot into another—
> suppose you thus successfully deal with a number of
> such instrumentalities; do you think these evils dis-
> appear absolutely? To see that they have not, you
> have to ask—Whence comes the official apparatus?
> What defrays the cost of working it? Who supplies
> the necessaries of life to its members through all
> their gradations and rank?

Spencer concludes,

> When it is thus seen that the evils are not removed,
> but at best only redistributed, and that the question
> in any case is whether redistribution, even if practi-
> cable, is desirable; it will be seen that the "must-do-
> something" plea is quite insufficient. (Spencer,
> 1896c:22-23)

The society which entails an equality of freedom is the most
egalitarian and progressive which can exist. Any attempt to
deny this leads to increasing misery for everyone.

This does not mean that the liberal goals of equality are forever relinquished. Rather, they would be postponed to a later
time when the contradiction between liberal values and social
reality would be overcome through the process of evolution
(Lakoff, 1964:145). Much like the meritocrats, there is a fundamental belief that if society is left alone by government most
of the major inequities will be resolved. For this reason, Spencer
opposed any and all governmental action including the inspection of gas works, prevention of child labor, laws enforcing
vaccination, taxing for drainage and irrigation, regulation of
safety in coal mines, checking the wholesomeness of food, a
public libraries tax "by which a majority can tax a minority
for their books," ad infinitum. Therefore, the state of
society which avoids government involvement is the preferred
position. Each deviance is dangerous and it is ultimately something society must be protected against. As Spencer writes:

> The function of Liberalism in the past is that of
> putting a limit to the powers of kings. The function
> of true Liberalism in the future will be that of put-
> ting a limit to the powers of Parliaments. (Spencer,
> 1896b:411)

Implications

The above discussion has not proved that Spencer's ideas
and the new conservatives' perspectives are identical. However,

their root-metaphors are undeniably similar. The key to this relationship is their view of equality and its interpretation in the libertarian tradition. Contrary to "nominal" claims, the new conservatives are liberals and are reflective of the kinds of divisions in liberalism which occurred in the late nineteenth century (Polanyi, 1975:ch. 12). It is a question of equality, and types of economic decisions involved in equality, to which is owed this hydralike phenomenon in liberalism. Lakoff credits this schism to epistemology (Lakoff, 1964:143), but our previous discussion appears to place it much deeper. Liberalism is an ideology which works best when society is under minimal strain (Dolbeare and Dolbeare, 1971). It tends to destroy itself on the altar of equality when the system it has fostered is critically assaulted. This is why Bell, and other new conservatives, or old libertarians, must cling so desperately to the defense of institutions. Once the ideology of liberalism is demonstrated as idealistic, or a cynical camouflage, refuge can only be found in structure—not function.

The root-metaphor of the new conservatives and Spencer has its origins in antiquity. As Pepper has pointed out, it probably began with the Milesian theory and one of its first proponents, Thales. For him, all things were water and he utilized this commonsense fact to explain life. This was followed by the addition of the qualities or development (change or "shaking out") of the substance of water (*apeiron*) by Anaximander and the characteristics of water—solid, liquid, gas—by Anaximenes. Thus, the theory would have (1) a generating substance, (2) principles, of change, and (3) generating substances produced by (1) through (2).

This does not give one a very adequate analytical perspective because it lacks scope. That is, there are too many anomalies that cannot be accounted for. If these crop up they are merely dismissed as unimportant or distorted to fit the theory. Inadequacies are seldom acknowledged and can never be corrected.[13] Spencer and the new conservatives have pushed their root-metaphor beyond what it can support:

> So we suggested that Anaximenes and Empedocles represented the generating substance theory at the

height of its Greek development. It is always possible that a theory may develop farther than the
best statement we have of it. In a sense, Herbert
Spencer's statement was a development beyond the
Greek. It was a development, however, chiefly in
respect to the vast accumulation of factual detail
over what the Greeks had, and hardly a development
at all in respect to the refinement of the categories.
It is the latter sort of development we chiefly have
in mind when we speak of the development of a
world hypothesis. For its adequacy depends on its
potentialities of description and explanation rather
than upon the accumulation of actual description,
though its power of description is never fully known
short of actual performances. (Pepper, 1966:97)

The new conservatives have also developed a plethora of facts
but have offered little in explanation. Though the meritocrats
do not consciously (at least to my knowledge) use evolutionism,
they do ascribe to its ultimate view of equality and its root-
metaphor. Where they are persuasive in facticity, they are bar-
ren in insight.

It is not too gross a statement to suggest that the meritocratic
position is truly an ideology of knowledge. The movement is a
natural outgrowth of the weaknesses inherent in liberalism and
certainly had a parallel development to the evolutionism of
Herbert Spencer. The essential thrust of this essay is that the
analytical position of the meritocrats is not an epistemological
question, but an ethical decision. It requires conscious value
assumptions which subsume the interpretation of "fact." In
this sense, I must agree with Paul Ricoeur: our understanding
of history creates our reality (Ricoeur, 1969). This created
reality, when consciously determined and ethically entailed,
must be considered an ideological perspective. This is not rigorous application of the term *ideology*, but there is certainly no
other term which adequately describes the use of an ethical
position to determine not only the analytical perspective, but
also its outcome. Let me be quite clear about my accusation
here: this type of ideology of knowledge can create the situa-

tions it describes simply because public policy is made in accordance with a stagnant world view.

It was Isaiah Berlin who suggested that Hegel had demonstrated that "great liberating ideas . . . inevitably turn into suffocating straight-jackets" (Bernstein, 1976:59). As has been demonstrated, the root-metaphor of the meritocrats has significant limitations which were first recognized in Greek thought. The new conservatives do present both a historical analogue and a comforting philosophical position for a people and government tired of confronting themselves. This is a movement trying to create history "as it was" for a nation dreadfully afraid of history "as it is."

The implications of a dominant meritocrat paradigm in the social sciences are evident. Some social science authors have already adopted it as the only analytical perspective. Because of its root-metaphor, it naturally erodes any contemporary conception of equality and logically ends with racism, class distinctions, and a justification of the status quo at any cost. Do I go too far? Surely it is a totally undeserved criticism of Professors Bell, Lipset, Kristol, et al. to accuse them of racism. However, the analytical perspective which they advocate, steeped in their view that man is inherently unequal, naturally leads to a "trickle-me-down" phenomenon in the social sciences. The avenues which are opened inevitably lead the students of the meritocrat professoriate to do "research" which is highly suggestive of this elitist perspective. This is in turn picked up by noted textbook writers and translated as givens to undergraduates, as well as the media. Again this is communicated to the offices of government, in whatever form, and becomes policy based on social scientific fact. In other words, social science is a dangerous business. Plato warned us of the poets who poison the minds of men with pernicious beliefs. His point was not made to induce censorship, but to see the danger of avoiding the ethical implications in our actions. The meritocrats resurrect a cause celèbre against equality which is deeply rooted in the American tradition. Yet few have criticized their ethical choice.

This should not be taken as a comfort to those who would prefer the Marxist perspective of political man. It only means that social scientists have an ethical decision to make.

Conclusion

It is one of those strange ironies of history that Karl Marx and Herbert Spencer are buried across the path from each other in London's Highgate Cemetery. This irony is the representation of the major conflict between root-metaphors in modern social science. And for this reason I would suggest that most of modern social science surrounds the question of equality. Spencer was much more open about his ethical values and, in an interesting way, provided a better justification for his position than the new conservatives. Contemporary society does not have the ability to judge virtue—as the Greeks—or view the morality of man—as Hobbes and Rousseau. Most questions in modern social science can be ultimately understood in terms of equality.[14] When confronted with contradictions, liberalism must turn either to the root-metaphor of Spencer or of Marx. This spectacular limitation has a great deal to do with our inability to do social or political analysis absent from an overriding view of equality. With this in mind, most research is premised on hidden ethical judgments, and little is conducted which can be considered insightful epistemological work. Richard Bernstein (1976:59) has admirably demonstrated this limitation.

He argues that there are two other paths open to social scientists: language analysis and phenomenology. The former is limited both in terms of applicability and the inability to examine its own premises. Phenomenology is certainly the most promising. It is self-conscious of its perspectives and "does not establish the ontological primacy of the manifest image or the *Lebenswelt* . . . [and] must also bracket the *Lebenswelt* if we are to understand its structures of meaning" (Bernstein, 1976:129). However, most social scientists are simply unable to make the necessary epistemological shift entailed in phenomenology from questions of reason to perspectives of understanding. At least for the time being, we must accept the inability of social science to engage itself in analysis and emphasize the ethical quandaries entailed in the creation of histories through notions of equality.

Social science in the 1970s is caught up in turmoil. Few can fully grasp its implications. If any credence is given to the analysis herein, a fundamental reevaluation of social scientific work is necessary in order to view its ethical impact. For we are engaged in viewing the politics of social science.

Notes

1. For example, this argument is accepted by Christenson (1976: chapter 4, section 5), which is the most popular-selling political science book by the publisher. It is also given as a paradox which can only be resolved in favor of inequality (Rein, 1976:173-84).

2. As Professor Herrnstein (1971b:110) points out: "as my article recounted, the greater the opportunities for social mobility and for equal education, the greater the social stratification according to genealogical factors, like inherited intelligence. This does not make me disfavor social mobility, nor do I welcome hereditary class. But whether we like it or not the former will create the latter." See Jerry Hirsch (1975a).

3. One of the best overall critiques is Ryan (1976), especially the appendices. Also Coser and Howe (1977) has a series of penetrating essays on the subject.

4. Critiques of this position can be found in Leon Kamin (1974) as well as the collection of articles in the *International Journal of Mental Health* (Hirsch et al., 1975).

5. Jensen (1969:55). The recent revelations about Sir Cyril Burt, who was most responsible for twin research, have clouded most of these finding Sir Cyril Burt suggested that he "regarded the actual data as merely an incidental backdrop for the illustration of the theoretical issues" (Hirsch, 1976). Herrnstein has argued that there is independent validity for the argument, but nonetheless all of the findings of the twin studies have been called into question.

6. This quote follows the statement on "the level of intelligence which is very, very common among Spanish-Indian and Mexican families of the Southwest and also among Negroes. Their dullness seems to be racial, or at least inherent in the family stocks from which they come. The fact one meets this type with such extraordinary frequency among Indians, Mexicans, and Negroes suggests quite forcibly that the whole question of racial differences in mental traits will have to be taken up anew and by experimental methods. The writer predicts that when this is done there will be discovered enormously significant racial differences in general intelligence, differences which cannot be wiped out by any scheme of mental culture" (Terman, 1916:91-92).

7. This approach is consistent from Seymour Martin Lipset (1960) through more modern thesis such as those put forth by Thomas Sowell (1976).

8. Nisbet (1974:106). Although I must admit the cuteness of Nisbet's reference to "the intellectuals," thereby—and somehow—distinguishing himself, I do feel it is more appropriate to take issue with his last statement. An argument can be made for calling "the politics of virtue from Plato to Rousseau" antidemocratic (see Nisbet, 1973, or Popper, 1950), however, it is important to note that arguments as to the democratic nature of virtuous politics have been made from Plato (e.g., Wild, 1953) to Rousseau (Chapman, 1956; Berman, 1970).

9. Cf. Andreski (1975), passim, and Carneiro (1968:121-28; 1973: 77-95).

10. This is not to suggest that Spencer does not. He clearly points out in Spencer (1896d) his four types of evolution. The question is whether these are distinctions in form or context.

11. Spencer (1896a:ch. XX). Also see Christenson (1974), especially his argument on women's rights.

12. Again this must be distinguished from the meritocratic argument which assumes that society indeed distributes intelligence with those with the greatest intelligence receiving the greatest rewards. There is a fundamental assumption that individuals are granted positions on the sole criterion of merit. Logically, those who are not at Harvard and are at Parsons College are naturally inferior. Does this also make Professor Bell more intelligent than the rest of us? *Mea culpa!* See Bell (1972:35).

13. "It is periodically revived in practically pure form, but always by men of relatively small caliber. It was revived by Bernadino Telesio in the sixteenth century and by Buchner, Haeckel, and Herbert Spencer in the nineteenth" (Pepper, 1966:93).

14. For example, Rawls (1971) has admirably demonstrated—though this was not his expressed purpose—that the question of justice is equality.

5

The Liberal Conception of Equal Opportunity and Its Egalitarian Critics

ROBERT L. SIMON

Equal opportunity long has been and still remains one of the guiding ideals of the liberal tradition. In its name, hereditary aristocracies have been overturned and race and sex discrimination condemned. At least as it has been interpreted within the liberal tradition, equality of opportunity has stood for a repudiation of caste and an affirmation of each individual's right to go as far as his or her talents allow.

Historically the doctrine of equality of opportunity frequentl῾ has been cited in support of distribution of scarce positions and resources on a competitive meritocratic basis. Competition would be unfair if the dice were loaded in advance so that some were virtually assured of winning and others of losing before the contest had even begun. Conversely, competition frequently is regarded as fair only when it is held under conditions of equal opportunity. Accordingly, proponents of the liberal ideal of equal opportunity generally have supported such measures as inheritance taxes and have been concerned with using the educational system to counteract the effects of unfavorable home environments or invidious discrimination. Fair competition and meritocratic distribution presuppose evaluation only according to each individual's relevant talents and abilities. Relevant talents and abilities in turn can most fairly be evaluated in a society in which equality of opportunity exists.

However, equal opportunity applied in a competitive framework need not result in equal outcomes. For example, if two baseball players of unequal ability and motivation compete for a position on a team under conditions of equal opportunity, the player with the greater ability and motivation will make the team. It is this aspect of equal opportunity—or the liberal interpretation of equal opportunity—that has aroused criticism from more stringent egalitarians. Such concern for inequality of result was eloquently expressed by Tawney, who suggested that

> It is possible that intelligent tadpoles reconcile them-
> selves to the inconveniences of their position by re-
> flecting that, though most of them will live and die
> as tadpoles, and nothing more, the more fortunate
> of the species will one day shed their tails, distend
> their mouths and stomachs, hop nimbly on to dry
> land, and croak addresses to their former friends on
> the virtues by means of which tadpoles of character
> and capacity can rise to be frogs. (Tawney, 1929:127)

This does not mean that egalitarian critics have abandoned the ideal of equal opportunity. Many instead have tried to divorce it from its competitive meritocratic foundations and supply more egalitarian interpretations. John H. Schaar, for example, has written that

> The equal opportunity principle is certainly not
> without value. Stripped of its antagonistic and
> inequalitarian overtones, the formula can be used
> to express the fundamental proposition that no
> member of the community should be denied the
> basic conditions necessary for the fullest participa-
> tion in the common life. (Schaar, 1967:242)

Equality of opportunity, then, has become a contested notion with opposing ideologies each trying to show that their interpretation is the "true" understanding of the ideal. Accordingly, it is crucial to distinguish different conceptions of equal opportunity. Otherwise what really is at stake may go unperceived because of unreflective acceptance of the mere formula of "equal opportunity."

As I have suggested, one conception of equal opportunity—perhaps the standard conception in the liberal tradition—is both individualistic and meritocratic or competitive in character. According to this conception, equal opportunity is a relationship holding among individuals when and only when outcomes are determined in a nondiscriminatory fashion.[1] Individuals of equal ability and motivation should do equally well, regardless of such factors as sex, race, or creed. Persons are judged on relevant factors alone. On this conception, scarce resources may be distributed competitively, but the competition must be fair or nondiscriminatory in character.

It is this competitive aspect of what I will call the liberal or meritocratic conception of equal opportunity that has troubled egalitarian critics such as Schaar. Realizing that even fair competition yields unequal results for unequal competitors, the egalitarian critics have turned against the meritocratic conception of equal opportunity. In what follows, significant egalitarian criticisms of equal opportunity will be explored. The specific objections to be examined are those developed by John Schaar in his important essay "Equality of Opportunity Beyond," since this essay is especially successful in articulating the concerns of the egalitarian critics. An examination of these criticisms will put us in a better position to determine if the meritocratic conception is defensible, whether it needs modification or whether it should be abandoned entirely.

The Critique of Liberal Equality of Opportunity

Equal Opportunity as Disguised Caste

The meritocratic or liberal conception of equality of opportunity originally was employed as a weapon against unjust discrimination and ascriptive hierarchies. It was held to be unfair that an individual's role was fixed because of the class or group into which he was born, his parents' status, fortune, or position, or because the individual himself possessed certain immutable physical characteristics. Rather, each individual was to be judged by his or her talents and abilities alone. Social mobility was to replace hierarchy, and competition rather than ascription was to be the order of the day.

However, according to egalitarian critics of the liberal conception, the view that implementation of equal opportunity is liberating is itself a deception. True, competitive distribution under conditions of equal opportunity does avoid ascriptive sorting by race, sex, and the like. But, according to the critics, it simply replaces the old invidious hierarchy with a new and no less invidious hierarchy of its own.

John Schaar (1967:233), for example, makes this point when he argues that the doctrine of meritocratic equal opportunity "removes the question of how men should be treated from the realm of human responsibility and returns it to 'nature.' What is so generous about telling a man he can go as far as his talents will take him when his talents are meager?" Competitive outcomes will be determined by such initial factors as the distribution of favorable early environments or perhaps a favorable draw in the genetic lottery, factors for which individual agents are not themselves responsible. No one chooses either his parents or his genes.

This argument might be more formally stated as follows:

(1) The more conditions of equal opportunity are implemented, the more the outcome of competition will depend upon possession of relevant talents and abilities.

(2) But possession of relevant talents and abilities is itself determined by factors beyond the individual agent's control, namely, distribution of favorable early environments and the genetic lottery.

(3) Therefore, the more conditions of equal opportunity are implemented, the more outcomes will be determined by factors beyond the agent's control, namely, the distribution of favorable early environments and the genetic lottery.

Moreover, it might appear that the more early home environments are equalized, the greater role genetic differences will have in determining outcomes.[2] Hence, Schaar remarks that rather than eliminating ascriptive hierarchies, the meritocratic doctrine of equal opportunity simply substitutes a natural ascriptive hierarchy for a social one.

This objection, it should be noted, applied to the competitive conception whether or not competitive outcomes are determine by achievement or by exertion of effort. For our capacity to exert effort and our motivation may be just as influenced by factors outside our control as our possession of certain abilities or talents.[3]

What is the proper response to the objection? Apparently, it will not do to go on equalizing environments, for, according t the critics, this will just allow for more and more genetic determination of outcomes. Indeed, even if all infants were raised in identical environments, different individuals will react different to the same circumstances, some more advantageously than others.[4] The problem of hierarchy would not have been remove

Perhaps what is needed, one might think, is not the total equa ization of all environments, with its disastrous impact upon the institution of the family, but rather a handicapping system such that those with greater natural capacities are held back in relatio to those with less. Ability would be equalized.

However, even if this rather impractical proposal could be carried out, results would then be determined by a meta-ability, namely, the ability to use one's other abilities most efficiently under pressure. And there is no reason to think this meta-ability is any less the product of the genetic and environmental lottery than any other ability or talent.

If the handicapping proposal is pressed to the limit, so that even our meta-abilities are canceled out, then we wind up in the situation described by John Wilson where we find

> two identical people playing tennis . . . but neither could ever win. The game would never get beyond 40-40 or perhaps since neither was the least bit better than the other, the very first rally of the match would be interminable, or at least last until both players dropped from exhaustion, presumably at the same time. (Wilson, 1966:73-74)

If one player did win, it could only be because of luck since every factor that could give one opponent an advantage over the other would have been neutralized by a handicap.

Surely what we have here is just another lottery. But it is hard to see why this kind of lottery is superior to the situation where ability, motivation, and character are rewarded. As Alan Goldman (1977:26) has argued, "The real contrast still reduces to that between rewarding chance versus rewarding effort and past and potential contribution."

But is Goldman's distinction between rewarding chance and rewarding other factors really viable? After all, it is just the egalitarian's position that those other factors are the result of chance themselves.

It is here, however, where it seems to me the egalitarian argument is misleading. Premise (2) of that argument states that possession of relevant talents and abilities is determined by factors beyond the individual's control. But what must be kept in mind here is the difference between the claim that we possess certain (perhaps all our) characteristics because of causal factors beyond our control and the different claim that social outcomes ought not to be affected by those characteristics. Since the second claim does not follow from the first, the implications of the egalitarian argument may not be as hostile to the meritocratic conception of equal opportunity as first appears.

It is frequently argued in defense of the meritocratic/liberal conception that it best promotes productive efficiency. What is not so often noticed is that this is not simply a quantitative point to the effect that more is better. For greater efficiency tends—when certain other conditions to be discussed later are also satisfied—to give people better chances to satisfy their desires and greater options about what sorts of desires to implement. As we will see, it puts greater control of people's lives in their own hands than would otherwise be the case, thereby broadening the realm in which we can function as autonomous agents.[5]

Perhaps the following hypothetical but I think illuminating example will be of help here. Imagine a small community of scholars and students that is joined by a new member. This new individual, whom we can call Jones, is a particularly interesting, acute, and stimulating instructor. Soon, more and more students are attending his lectures and less and less are attending those of other scholars. Other scholarly communities hear of Jones and try to induce him to join them. The students in the original community do not want to lose Jones and offer him counter-inducements to stay. Colleagues who admire Jones and wish to emulate him (or alternately want to receive the same induce-ments and counterinducements themselves) adopt his methods of teaching or try to develop better methods of their own. In short, a perhaps embryonic but nevertheless real form of com-petition, with correlative inequality of result, develops.

If such competition had not been allowed—perhaps by requir-ing that each instructor have an equal amount of students—the price in terms of respect for persons would have been severe. From the perspective of the students, their critical choices would have been allowed to affect outcomes. Many would have had to attend certain classes against their will. Similarly, Jones's life would have been relatively unaffected by his choices and abilities. He would have come out the same whether or not he developed his teaching skills, no matter how hard he tried to be a superb teacher. The competition in question not only promotes effi-ciency in the sense of providing an incentive for superior teach-ing. More fundamentally, it arises from the choices of individuals as to how their lives are to be led!

Perhaps the ethically most important defense of competition is that it places responsibility for outcomes in the hands of af-fected persons. It assigns them the status of choosing autono-mous agents.[6] Indeed, it is hard to see how competition could be eliminated without severe restrictions on the ability of indi-viduals to choose which activities to enter or on the freedom to admire or criticize the performance of others and offer op-portunities for others to repeat those performances.

This does not imply that choice never should be interfered with. Even in our hypothetical example, it is arguable that students ought to be required to take certain courses which are essential liberal arts education. But even here, the point—or at least a significant point—of requiring certain courses is to place students who have taken them in a better position to make crit-ical informed choices in the future. When choice is restricted, the more such restrictions promote our abilities to choose more intelligently and capably in the future, the easier they are to justify.

Accordingly, the charge that the liberal conception of equal opportunity really is only an ideological mask for a new invidious hierarchy is itself open to strong objection. Even if it is true that our character, our values, our skills, and our abilities are all ultimately the product of causal factors beyond our control, it does not follow that they ought to be disregarded. Indeed, in any system some person's choices are going to affect outcomes. Equal opportunity broadens the class of people whose choices have such weight by prohibiting discrimination on morally suspect grounds. Unlike caste systems based on race, sex, or ethnicity, it respects persons by requiring that outcomes be significantly affected by the character, choices, and skills of individual men and women themselves.

Equal Opportunity and Unequal Result

It is true, however, that pervasive competitive practices, even under conditions of equal opportunity, may lead to extensive inequalities of condition. Thus, Schaar (1967:237) charges that behind the meritocratic conception of equal opportunity lies a view "of human relations as a contest in which each man competes with his fellows for scarce goods, a contest in which there is never enough for everybody, and where one man's gain is usually another's loss."

Before dealing directly with this point, an important but logically prior matter needs to be dealt with first. What may trouble many thoughtful people about competition in this society is that it is biased by the effects of past discrimination, deprivation, and injustice. In particular, the legacy of slavery, segregation, and more covert forms of discrimination may especially disadvantage victimized minority groups. Surely open-ended competition under such conditions is unfair since some but not others labor under unjustly imposed competitive handicaps.

However, it certainly would seem that adherents of equal opportunity, in the name of that very ideal itself, are committed to some form of corrective or affirmative action. For according to the very conception of competitive equal opportunity, outcomes are to be determined by possession of relevant characteristics and qualities, not by prejudice or discrimination. Where the latter are likely to effect results, there is an obligation, based on equal opportunity itself, to try to minimize the impact of such injustice.

This does not mean, of course, that all forms of affirmative action are equally legitimate. At least some of Alan Bakke's defenders claim not to be opponents of affirmative action itself, but only of affirmative action programs that distribute benefits along lines of race, sex, or ethnicity.[7] The issue of what an acceptable form of affirmative action would be like raises issues that go far beyond the scope of this paper. What I suggest, however, is that commitment to equal opportunity may well require commitment to some form of affirmative action but that it remains debatable whether affirmative action should be on behalf of all disadvantaged or victimized individuals, regardless of group membership, or on the contrary should be only for members of especially victimized minority groups.[8]

Be that as it may, doesn't Schaar's major point remain? Isn't it true that competition, even under fair background conditions, may yield significant inequality of result? And isn't that a point against it?

These charges should not be answered directly, for they oversimplify what is at stake. Instead, a number of distinctions need to be made.

First, we need to be clearer about whether equal opportunity is to refer to a person's chances to use his abilities or to an individual's chances to acquire abilities and talents, indeed to an individual's chances to become a certain sort of person.[9]

An example from an essay by Bernard Williams illustrates the distinction. Williams (1971:132) asks us to imagine a society in which "great prestige is attached to membership of a warrior class, the duties of which require great physical strength." In the past, only members of the wealthiest class served as warriors, so in effect the warrior class had constituted a hereditary aristocracy. But then

> egalitarian reformers achieve a change in the rules,
> by which warriors are recruited from all sections

of society, on the results of a suitable competition.
The effect of this, however, is that the wealthy
families still provide virtually all the warriors, because the rest of the populace is so undernourished
by reason of poverty that their physical strength is
inferior to that of the wealthy. . . . The reformers
protest that equality of opportunity has not really
been achieved; the wealthy reply that it has, and that
the poor now have the opportunity of becoming warriors—it is just bad luck that their characteristics are
such that they do not pass the test. "We are not,"
they might say, "excluding anyone for being poor;
we exclude people for being weak, and it is unfortunate that those who are poor are also weak."
(Williams, 1971:132)

Such a response is not likely to strike many of us as satis-
factory. This is because competitive equality of opportunity
seems to be simply an ideological prop for the established order.
Some do not have an equal opportunity to develop their capacities and abilities—or even to acquire the motivation to develop
the capacities and talents—which society values.

Once again, this need not mean that each individual has an
equal probability of developing in a particular way. Such an
ideal might be achievable only by a society of genetically identical individuals raised in identical environments, an ideal that
might well be as indefensible as it is unattainable in practice.
What it does seem to require, however, is that each individual
have equal access to the basic goods and services necessary for
development of his or her character, capacities, and abilities.
By basic goods and services, I mean such things as an adequate
diet, decent clothing and shelter, and decent educational and
health facilities. These goods and services are basic in the sense
that without them, individuals cannot develop their capacities
and abilities, or even acquire the motivation to want those
capacities and abilities to come to fruition. Accordingly, the
competitive conception of equal opportunity seems defensible
only when it encompasses a more basic developmental conception of equal opportunity. This developmental conception requires that every individual have the same chances of access to

those basic goods without which full development as a person is impossible.

Of equal importance, commitment to equal opportunity need not be the only star in one's firmament of social justice. On the contrary, commitment to equal opportunity is compatible with allegiance to other principles of social justices as well. For example, one might hold that offices and positions be distributed competitively but that unequal rewards attached to those positions be allowed only if the worst-off group in society would be even worse off were the inequality eliminated.[10] Or one might hold on independent grounds of social justice that no one be allowed to make less than one-half the median income in one's society. Equal opportunity is only one component in a "total social justice package." Accordingly, commitment to equal opportunity is compatible with a wide range of positions on the scope and limits of permissible inequalities in other areas.[11]

Once the liberal meritocratic conception of equal opportunity is understood as encompassing developmental equality of opportunity and as simply one element in an overall conception of social justice, it is compatible with significant limits on permissible inequality of condition. At a minimum, the distribution of goods and services may not be so unequal as to deprive anyone of access to adequate food, shelter, and clothing or access to decent health and educational facilities. In practice, this may require substantial restrictions on many inequalities in our own society. For example, it may require abolition of systems of school financing which allow wealthier school districts to spend far more per pupil than poorer districts are able to spend.[12] The extent of permissible inequality of result will often prove controversial, and implementation of the developmental ideal may itself raise practical and ethical issues. It is doubtful, however, whether the advocate of competitive equality of opportunity is committed to whatever degree of inequality of result the competitive process dictates.

Equal Opportunity as an Apology for the Status Quo

It may be conceded at this point that equal opportunity does indeed allow for outcomes to be significantly affected by individual choice, but only within limits. For what equal opportunity really ensures is only the ability to compete within an already established framework of social values. This point is forcefully presented by John Schaar. According to Schaar,

> the equal opportunity formula must be revised to read: equality of opportunity for all to develop those talents which are highly valued by a given people at a given time. . . . When put in this way, it becomes clear that the commitment to the formula implies prior acceptance of an already established social-moral order. Thus, the doctrine is, indirectly, very conservative. (Schaar, 1967:230)

What Professor Schaar may not have noticed is that exactly the same charge could be made against democracy:

> the democratic formula must be revised to read: everyone has an equal chance to participate as an equal in an election, the result of which will be the selection of policies or persons most highly valued by a given people at a given time. When put in this way, it becomes clear that commitment to the formula implies prior acceptance of an already established social-moral order. Thus, the doctrine is, indirectly, very conservative.

The mistake here, I suggest, is to equate proposition (1) below with (2).

(1)   A democratic election expresses the values to which people are committed at the time of the election.
(2)   These values are supportive of the status quo.

(1) is a truism. If people are given a choice, then, necessarily, the values they hold at the time of the choice will influence outcomes. It does not follow, however, that the values in question will be conservative, in the sense of endorsing things as they now are. Similarly, competitive practices may reflect currently held values in a variety of ways. For example, people may choose

to compete in certain areas rather than others because success in those areas is valued more highly than in others. It does not follow, however, that such values must be conservative or that they are not open to change. The recent growth of interest in soccer in the United States illustrates, if only in a small way, that people's values are open to change and, in addition, that change itself can be stimulated by the competitive success of new values in driving out or modifying old ones.

Of course, Schaar need not be committed to the view that competitive systems are unchanging. Rather, he might simply argue that as a matter of empirical fact, competition, even under conditions of equal opportunity, tends to perpetuate established values.

However, the same point can be made with equal plausibility about democracy. But what are the alternatives? Surely, there is a strong case against disregarding individual preferences, both in the political arenas of the democracies and as they are expressed in competitive arenas. But once individual choice is accorded significant weight, inequalities naturally arise, whether through the competitive process of elections or through the kinds of selections ordinary people make in the course of running their lives. And it is just those choices, I have contended, from which inequality of condition arises.

Moreover, Professor Schaar has not shown that other systems tend to be less conservative than the ones he criticizes. Indeed, given his line of argument, there is no reason why they should be any more amenable to change. For somewhere in those systems, some individuals will have to make choices about how things are to go. But those individuals too have been "socialized" and so will choose in terms of their own values, values which presumably are in some harmony with those of the preexisting society around them. Why should their choices be any less conservative than those made under conditions of equal opportunity? At least under equal opportunities, proponents of new ideas may try, sometimes successfully, to overthrow established beliefs.

The charge that equality of opportunity is an inherently conservative ideal is, then, open to three lines of rebuttal. First, it may be argued that the charge rests on a confusion between

choices based on the values of individual persons at a given time
and choice based on values supportive of the status quo. The
mistake is to assume that the former are inevitably the latter
as well. Second, even if equality of opportunity tends to work
in a manner supportive of the status quo, this may be the price
one has to pay for respecting the choices of individual persons.
At least, it is not obvious that the latter should automatically
be disregarded so as to avoid the former. Finally, it is unclear
that a society dedicated to the ideal of equal opportunity will
be more conservative than other societies. Much will depend
on the ideals to which the others are dedicated. There is at
least some reason to think, though, that decisions in many
other kinds of social frameworks will be far more conservative
as those made within the framework of liberal equality of
opportunity.

The Value of the Liberal Conception

I have argued that contrary to its egalitarian critics, the liberal
meritocratic conception of equality of opportunity is not ascriptive, need not promote gross inequality of condition, and is
not necessarily supportive of the status quo. On the contrary,
it provides the rationale for condemnation of many unjust and
outrageous inequalities that exclude all too many individuals
from full participation in our society.

However, the liberal conception does classify inequalities as
legitimate when they arise from the unprejudiced, uncoerced
choices of individuals, under background conditions of developmental equality of opportunity and nonviolation of other principles of social justice. Accordingly, the liberal conception does
not automatically associate inequality with injustice but, on
the contrary, specifies conditions under which inequalities are
just and fair.

This point is missed by some social critics. Christopher Jencks
(1972) has suggested that liberals of the 1960s saw equalization
of educational opportunity as a means of equalizing incomes
later in life. He then suggests that this strategy is mistaken since
equalization of educational opportunities will not radically
reduce income inequality (Jencks, 1972:8-9). What this analysis

ignores is that equal educational opportunity need not be viewed as a means of securing greater equality of income or wealth. Rather, on the liberal conception, it may be seen as a condition legitimatizing inequality of outcome.[13] The issue of whether or not incomes should be made more equal indeed is an important one. But it only confuses what is at stake to simply assume that the point of equal opportunity is to promote equal result when in fact it may promote conditions under which unequal results can be fairly arrived at.

The importance of this point can be best brought out by considering an alternate conception of equal opportunity based on the idea of a lottery. According to the lottery conception, when goods to be distributed are scarce, the only fair way to distribute them is by chance. Although no actual theorist may have actually advocated such an idea, at least in so extreme a form, it will prove useful to consider the presumption behind this hypothetical ideal, namely, that social institutions ought to ensure each individual an equal probability of benefiting where distribution of scarce but important goods and services is at issue.

Suppose it is claimed, for example, that scarce places in law or medical school should be distributed by lot. To be eligible for the lottery, one must be academically qualified. But since the qualified vastly outnumber the slots available, only lottery winners actually will be able to attend. The alleged advantage of this procedure is that purely arbitrary differences among candidates will not affect outcomes. Although Jane may score twenty points higher than John on the LSAT, and on that account perhaps rank ahead of John under normal admission procedures, the difference may be due to Jane's more favorable home environment which she did nothing to deserve. Given that Jane and John are both academically qualified, Jane's higher score is irrelevant to the outcome of the lottery. Both Jane and John have an equal chance of being winners.

However, the disadvantages of the lottery proposal are great indeed. To see this, consider the case of Bill and Sue. Bill is naturally intelligent, has had many advantages, and is able to do acceptable academic work with relative ease. On the other hand, Sue, who may be no more intelligent than Bill, has overcome severe disadvantages by hard work. She is intellectually curious, highly motivated, and works especially hard to overcome her weaknesses. As a result of superior effort in the face of disadvantage, her academic record is superior to Bill's.

What is the impact of the lottery system here? First, it forces us to ignore the important differences between Bill's and Sue's performances. These differences not only are relevant to rational prediction about their academic success in graduate school and their performance as future physicians or lawyers. Of at least equal importance, they are also relevant to our estimation of Bill and Sue as persons. The implementation of the lottery renders us incapable of allowing that estimation to affect outcomes. Our values are made impotent; they have no effect upon who is selected and who is not. Second, the lottery serves as an incentive for others to act like Bill. Why strive for the best within us when by merely being qualified we have as good a chance as anyone else? Conversely, an extra burden is imposed on people like Sue. How many will hold out for how long against taunts—taunts that reflect the truth—that hard work is foolish and inefficient? (It will do no good to object that hard work will pay off later, for by the logic of the lottery conception, other important benefits and burdens should be distributed by lottery too!) Finally, society is prohibited from selecting the best qualified candidates. Perhaps those most adversely affected by the lottery are those consumers, including many disadvantaged and deprived individuals, who get less efficient services than would otherwise be the case.

Lotteries may well be appropriate when distribution of scarce basic goods or lifesaving resources is at issue. In such contexts, it is unclear that individual differences ought to affect outcomes. We are inclined to say that all individuals, simply in virtue of their status as persons, ought to have equal chances. That is why it is appropriate to define developmental equality of opportunity in terms of equal probabilities.

But to extend the lottery ideal beyond such a sphere and elevate it to the status of a fundamental distributive principle is unwise. It is to ignore those facets of the individual most central to our status as persons. It is to say that our character, capacities, choices, and actions should not affect outcomes.

Conversely, it is those central features or our status as persons that the liberal conception of equal opportunity puts in the limelight. For equal opportunity, on this conception, is just the equal chance to influence outcomes through our choices, character, and capacities expressed in action. It is this that constitutes its greatest value.

Notes

1. What counts as nondiscrimination will often be controversial. For my purposes, a procedure is nondiscriminating if it does not distinguish invidiously among individuals according to such factors as their race, sex, religion, or ethnic background. Exactly what other factors ought to be included is unclear, and controversy frequently will break out over inclusion or exclusion of particular items. Often, what is discrimination or not will depend on context. For example, it may be legitimate to consider attractiveness when hiring a model but not when hiring a salesperson.

2. On the principle that the less environments of members of a given population differ, the more variance among them is due to differences in genetic endowment. The belief that liberal emphasis on equality of opportunity only promotes the creation of a genetically grounded meritocracy has been defended by Herrnstein (1971).

3. Thus, John Rawls (1971:104) writes, "The assertion that a man deserves the superior character that enables him to make the effort to cultivate his abilities is equally problematic; for his character depends in large part upon fortunate family and social circumstances for which he can claim no credit."

4. Perhaps each individual should be placed in an environment uniquel suited to develop his or her talent to the fullest. Unfortunately, this proposal runs into many difficulties. Among the most significant are (a) the difficulty in discovering what an individual's talents are at an early enough age to affect their development significantly; (b) the difficulty of deciding which talents to focus on; and (c) the problem for political liberty raised by the fact that the individual might not have selected just those talents for emphasis, given the opportunity to make a rational choice.

5. I have argued along somewhat different lines and at greater length for this thesis in "An Indirect Defense of the Merit Principle" (Simon 1978

6. This suggests that the issue of human responsibility is ultimately normative and profitably can be separated from metaphysical disputes about causal determinism of behavior.

7. See, for example, Bunzel (1977), "Bakke vs. the University of California," particularly p. 64.

8. For discussion of these and related questions, see the essays re-

printed in Cohen, Nagel, and Scanlon (1977), *Equality and Preferential Treatment.*

9. I am indebted to the useful discussion of a similar distinction by Frankel (1971) entitled "Equality of Opportunity."

10. Such a view is compatible with the position defended by Rawls (1971).

11. For a useful discussion of a similar point involving merit, see Daniels (1978), "Merit and Meritocracy."

12. I have discussed this issue in "Equal Opportunity and the Serrano Decision" (Simon 1973).

13. Here, I am following the discussion of Kavka in his "Equality in Education" (1976).

II

IMPLICATIONS FOR
SOCIAL POLICY

6

Biological Differences and Economic Equality: Race and Sex

MASAKO N. DARROUGH

Introduction

> We hold these truths to be self-evident, that all men are
> created equal.
>
> Declaration of Independence

It may also be self-evident that the notion of "equality" in the Declaration of Independence is not that of identity. That all men are created equal does not imply that all men are created identical physiologically or mentally. The new equality here entails "equal rights" to "life, liberty, and the pursuit of happiness." These fundamental rights are considered to remain constant, unaffected, and unchanged, even if people may differ from one another in shape, size, skill, and every conceivable attribute. Exactly how these rights translate into a social, political, and economic system is not self-evident and is a matter open to considerable debate and interpretation.

It may be said that an individual derives inalienable rights solely by being a member of a particular group. Membership alone is considered sufficient to ascribe the individual with fundamental rights. On the other hand, exactly who should constitute the group may not be so self-evident, and, in fact, the membership has changed significantly in the last two centuries. Indeed it was drastically expanded in 1863 (Emancipation Proclamation), 1920 (Women's Suffrage), and 1964 (Civil Rights

Act). However, at the time of the Declaration of Independence, the founding fathers meant precisely "men," and only men of a certain color.

Two sets of people who were denied membership in "the group of equals" on the basis of their biological differences were blacks and women. The rhetoric behind their exclusion has often sounded very similar. Bird (1971:110-11) suggested:

> Both women and blacks were held to be inferior in intelligence, incapable of genius, emotional, childlike, irresponsible, and sexually threatening. They are supposed to be all right in their place, and were presumed to prefer staying there. . . . Both were viewed as treacherous, wily, "intuitive," voluble, and proud of outwitting their menfolk or white folk.

These analogies and parallels between race and sex have been pointed out by many others (Montagu, 1963; Myrdal, 1944; Pole, 1978) as well. Montagu (1963:127) stated, in 1946, that "practically every one of the arguments used for racists to 'prove' the inferiority of this or that 'race' was not so long ago used by the anti-feminists to 'prove' the inferiority of the female as compared to the male."

Blacks and women have now gained "membership status" and are assured constitutional rights in many respects. In addition, over the last two decades there have been many significant social reforms in the realm of economic discrimination. These reforms deal with more practical issues in regard to an individual's basic rights in the marketplace. Discrimination according to race, sex, national origin, and religion is proclaimed to be in violation of one's rights. In most legislation, sex and race (as well as other categories) are treated as analogous (although sex was introduced into Title VII of the Civil Rights Act as a "joke" Frequently, women and nonwhites are lumped into a group called "minorities."

One purpose of this essay is to establish the significance of the analogies between race and sex and to investigate further the extent to which they are valid. It will be argued that, in fact, the analogies are not very extensive, and for that reason treating race and sex in the same manner is actually misleading and could be inefficient in formulating policies. We will do this by pointing out two important asymmetries existing between race and sex that are of crucial relevance in promoting some sort of economic equality in society.

This essay has evolved out of an attempt to investigate the broader topic of biological differences and economic equality. Our first question was: How are biological differences among human beings to be handled in looking at the issue of economic equality/inequality? Two subquestions will be addressed specifically. First, what biological differences are relevant to the organization of economic systems and, in particular, to the individual's role as an economic agent in a system? Second, precisely what is meant by the notion of economic equality/inequality, and how does it relate to these biological differences?

Although all individuals (except monozygotic twins) are different in genotype and phenotype, it is possible to group people into various categories according to biological characteristics such as height, blood type, race, and sex. In this essay, we will consider only differences of race and sex. Both sex and race are sociologically as well as biologically significant concepts in our society. What then is the significance of race or sex differences in discussing the question of equality in the economic dimension?

In order to come to grips with the question raised above, we will relate it to the problems a society such as ours may face in the relatively near future. We shall call it the "post-Bakke world." In other words, we will investigate the question for a particular type of social organization, rather than for an "ideal" society.

In the post-Bakke world, we assume the following:

(1)   There are differences among human beings, i.e., all human beings are not created equal, although equality and equity are considered desirable.

(2)   Some sort of meritocratic system is justified due to incentive consideration, although the degree of meritocracy (or how rewards are determined by merits) is a variable as a tool of social choice.

(3)   An approximately free market economy exists.

It is clear that in the post-Bakke world above, economic equality does not imply economic identity. Yet equality entails at least "equal opportunities" in the marketplace regardless of one's race or sex. The most difficult question to untangle here is what these "equal opportunities" imply if and when we are dealing with individuals with different attributes (biologically as well as others). The idea of "fair competition" or "fair game" among "equals" is appealing and consistent with American popular sentiment. But what is fair competition among "unequals"?

The first section of the essay will focus on the discussion of biological differences, in particular, race and sex. Of what importance to social scientists are biological differences among races and between sexes? Are these concepts biological and/or sociological?

Lessons from sociobiology are incorporated in the second section. What biological forces are operating on *Homo sapiens?* Is there "natural order" which may be determined by our biology and is truly beyond our control? The parallel between the models from sociobiology and economics is drawn, and its limitations for problems at hand are pointed out.

The third section will discuss asymmetries between race and sex differences in dealing with the question of equality. Two important asymmetries arise due to the organization of property rights in society and to the reproductive roles played by the individuals.

In the last section, we will attempt to incorporate these asymmetries in order to formulate economic and social policies. Given some kind of social objectives, which express commitment to "justice," "equality," and "equity," what are the policy implications of these asymmetries?

Biological Differences: What Difference Do They Make?

It is obvious from our casual observations that no human being is exactly the same as another. Even identical twins are usually distinguishable by sight when they are adults, although they may have looked like two peas in a pod when they were smaller. Apart from identical twins, most human beings are quite different from one another in terms of physical characteristics, personality, mental capacity, and attitudes. On the other hand, in the sense that all humans are clearly distinguishable from any other species of animals, they do share many things in common. In fact, all humans share the same genes (although humans share 99 percent of their genes with chimpanzees). The genes which separate *Homo sapiens* from the other animals may be what enable us to perform many activities which are characteristic only of humans, such as walking erect, communicating by using (sophisticated) language (Washburn, 1978), and engaging in sexual activities "throughout the calendar, if not around the clock" (Barash, 1977:296). Furthermore, all populations of humans are "inter-fertile."[2]

Thus it is clear that humans are alike in many aspects, separating us from other animals, in genotypic as well as phenotypic characteristics. Keeping this in mind, the next step is to investigate and evaluate the variations and differences within the species. The degree of variability in the gene configuration is crucial in the evolutionary process, since it provides more "adaptive" opportunities to a changing environment.

Suppose we define biological differences (as opposed to cultural differences) to mean all phenotypic distinctions (physiological, morphological, and behavioral) which are basically determined by genetic factors. Although genetic factors are inherited at birth, subsequent environmental contributions also influence the nature of phenotypic characteristics. (For example, an individual's height and weight could be influenced by, among other things, the diet, although something like blood type would not be.) However, since it is almost impossible to talk about all possible variations in the genotypes,[3] we often group the individuals into a smaller number of categories according to biological characteristics, such as blood type, type of ear wax, height, weight, skin color, skeletal structure, hair texture, sex, age, eye color, lip shape, and so forth. Some of these differences are not only intrinsically interesting; in some medical treatments an understanding of these variations is vital. Since particular clusters of certain biological characteristics are not randomly distributed, information on group characteristics and their frequency distribution may be very useful for diagnostic and policy purposes. Thus biologists and medical scientists, among others, are quite interested in these biological differences.

To social scientists, however, these biological differences present somewhat different problems. In general, we are not interested in the ear wax type. The differences are of interest to us if and only if they are socially significant, that is, if people assign some social significance to the particular biological difference in distributing scarce goods such as jobs, status, prestige, power, income, and so forth.

In this sense, looking at social organizations through particular blood types of people is probably futile and meaningless, unless one is interested in administering a blood bank.[4] Then what are more interesting biological differences in examining social and, in particular, economic equality? Three biological differences may be considered relevant to contemporary American society: race, sex, and age. In this essay, we will only discuss the first two, since age may be regarded as inherently different from race and sex. Typically, race and sex are considered mutually exclusive categories for any individual at all times.[5] However, the age of the same individual will be constantly changing. Therefore, the biological difference in age may be examined adequately simply by looking at the dynamic or lifetime profile of individuals.[6]

At this moment, although we are not too concerned with exactly how races are defined,[7] it is important to note that the concept of race is both biological (genetic) and sociological. (Of course, it is also evident that the concept of sex is based on biology.) Even if all humans may have evolved from a few people in the same geographic area, subsequent geographic dispersion and genetic drift, combined with natural selection, seem to have created groups of people with somewhat distinct biological characteristics. Although there is a considerable degree of diversity within each race, the frequency distribution of some characteristics are quite race specific. For example, the frequency distribution of ear wax type, fingerprints, blood type, skin color, lactose level, as well as incidences of some diseases (e.g., sickle-cell anemia) are often distinguishable by race.

Thus the average differences among races are relatively easily established and proven to be useful, particularly in the sphere of medicine. Yet it must be pointed out that these averages may be of no use in knowing the characteristics of a particular individual taken at random. Comparing two pairs taken randomly from two racial populations and examining the differences in the chemical structure of the proteins produced by these four individuals, both the intra and inter group differences are reported to be—on the average—about the same, with differences of 0.02 percent (Goldsby, 1977:63). Goldsby points out two important lessons: "First, there is considerable variation between human beings, whether they belong to the same racial group or not. Secondly, racial differences add very little to the variation that already exists between human beings."

Of course, some of these differences are polymorphic and not really measurable by a single continuous variable. Discussions such as the above, therefore, may not convincingly resolve issues on racism. If one is willing to assign a large enough weight to one characteristic, such as skin color, all other differences may be dismissed as trivial.[8]

Similar statistics on average differences and their frequency distributions are available for the differences in sex; however, the most important element in the concept of sex from the biological point of view is the dimorphic reproductive functions of the two sexes. The female reproductive functions, in particular, may not be limited to the acts of mating, being pregnant, and giving birth. Biologically, female bodies must develop to carry out these functions, and this may involve years before and after the actual production of offspring. At least at this moment, the biological functions in this aspect of males and females are not interchangeable. Females alone are capable of carrying the fetus, giving birth, and nursing the offspring.

In addition to this functional difference, we see some evidence of average differences in height, weight, physical strength, endurance, life expectancy, cerebral organization (McGlone, 1980), the frequency distribution of diseases, and illness between the two sexes.[9] Here again these average differences may not help in speculating about the characteristics of any particular individual taken at random.

Sociobiology and Culture

> The central question is whether or not the human species
> have entered a new domain of experience where general bio-
> logical laws will have only negligible relevance or have been
> abolished by the unique developmental advances achieved
> by mankind.
>
> Hirshleifer (1978:52)

One attempt to incorporate biology into the investigation of
social behavior of organisms is sociobiology, discussed extensively by Hamilton (1964), Wilson (1975, 1978), Barash (1977)
and others.[10] Sociobiology seeks to "develop general laws of the
evolution and biology of social behavior" of organisms and,
perhaps, to extend such laws to the study of human beings
(Washburn, 1978:406). It is based on the Darwinian notion of
natural selection which results in the differential survival of
organisms, selecting the so-called "fittest." The different adaptive behaviors of these organisms allow the forces of natural
selection to change the gene frequencies and, hence, evolution.
The genetic materials of the individual organisms, such as DNA
molecules, are "blindly programmed" or "coded" so that the
individuals behave in such a way as to maximize both the "reproductive and inclusive fitness" (Hamilton, 1964). The unit
of natural selection may be genes (Dawkins, 1976), an individual organism, a population or a species as a whole. Therefore,
if these organisms are to live in groups, their survival would be
enhanced by cooperation with one another. The individual organisms compete for scarce resources within the groups on the
one hand; however, they may engage in cooperative behaviors
which may have evolved as selectively more adaptive genetic
traits on the other. Therefore, the theory of kin selection (or
inclusive fitness), which explains altruistic behaviors, are integral parts of sociobiology.

Sociobiological models seem to fit the animal kingdom quite
well. After all, what else can we conjecture but that the animals
are programmed to maximize, if not consciously, their survival
(of their gene pool)? Those who do not will be selected against
eventually by nature and will not evolve subsequently. The

animals which do survive are the ones which are capable of behaving as if maximizing their reproductive fitness. The concept of evolution through natural selection in this sense is not necessarily loaded with any normative judgment, but does imply that the survivors are the ones who do matter eventually. Nonsurviving genes will not be represented. Evolution is ultimately opportunistic.

Since the model is that of constrained maximization, the basic sociobiological model of individual behavior—even though governed by biological rather than cultural forces—may be of particular interest to economists (Becker, 1976; Ghiselin, 1978; Gordon, 1979; Hirshleifer, 1978; Samuelson, 1978). Subject to both environmental (ecological) and biological (genetic) constraints, what is to be maximized here is the individual's "reproductive success." For example, the individual has control over kin selection, but the basic behavioral patterns may have been already "programmed" in the genes and cannot be altered at will. In this framework of optimizing behavior, both egoistic and "seemingly altruistic" behaviors are consistent, although "true altruism" may not have any foundation in sociobiology (Becker, 1977; Tullock, 1977).

When sociobiology is extended to human behavior, the basic premise concerning the nature of human beings is that they are essentially selfish. However, since they are merely the "survival machines" of the "selfish genes," altruistic behaviors may arise in order to perpetuate the genes even by sacrificing the organisms. In addition, altruistic behaviors of a reciprocal nature may prove to be "adaptive." Since people cannot live in isolation, it becomes beneficial to cooperate by engaging in reciprocally altruistic behaviors (Trivers, 1971). Humans can be basically individualistic and selfish, but nonselfish social behaviors may take place as the result of implicit "social contracts."

Perhaps this sort of selfish individual model fits the ideal of "economic man" or "rational man" very well. This may be the reason why economists spend a great deal of time attempting to show how atomistically perfect competition may achieve a Pareto optimum. In neoclassical economic models, however, what we have done is to respecify the objective function. Instead of maximizing inclusive fitness, individuals are said to be maximizing "utility" or "satisfaction" (which may be a function of the consumption of goods, leisure by the individual, by the others, and so on). Ordinarily the outlook is focused upon the individual's current or lifetime utility and not beyond. Herein lies the fundamental difference between sociobiology and economics.[11] Individuals are not assumed to be merely survival machines; instead they are "pleasure" machines.[12]

The extent to which sociological models are appropriate, useful, or misleading for human society is the subject of controversy (Montagu, 1980; Sahlins, 1976; Washburn, 1978). Initially, it is not clear what our objective function is or should be. The opponents of sociobiology offer "evidence" of human behavior in which a nonreproductive goal appears to be the objective. Sahlins (1976:36) argues that:

> for human beings, survival is not figured in terms of life and death or as the number of genes one transmits to succeeding generations. Humans do not perpetuate themselves as physical but as social beings. Death is not the end of a man, nor even of his reproductive ability. Men alone are immortal.

In this light, abortion, celibacy, birth control, incest, infanticide homosexuality, suicide, and child adoption appear to be contrary to reproductive goals.[13]

Second, it is not clear what a "kinship" relationship implies and how far it extends for human beings. In most societies, there is agreement that the family constitutes the basic unit of social organization. These families are related and interrelated to other families with differential genetic and social ties. In most societies, families are composed of individuals who are biologically related. On the other hand, Sahlins points out many cases in which blood relationship is neither necessary nor sufficient for family relationship. Adoption is a common practice. Here social contracts, not biology, define the kinship relationship. Of course, in kinship relations, the biological factor remains a determining factor. However, it may not be as essential as one may be led to believe. Indeed, it may be quite arbitrary in human social organizations.

The model of reproductive calculus illuminates an interesting aspect of altruism. Reciprocal altruism is based on a benefit-cost analysis where benefit to the benefactor is inversely related to the degree of kin relatedness. It is consistent with such phenomena as small-town mentality, provincialism, urban alienation, and nationalism. At the other end of the spectrum, human society is full of cases in which strong hatred, jealousy, and antagonism arise between parents and offspring, among siblings and close relatives. One needs only to recall Hamlet, Caligula, Elizabeth I, Cain and Abel to name just a few instances.

Barash (1977) argues that it is presumptuous and arrogant to dismiss biological and evolutionary forces in studying human behavior.[14] It is not merely a question of nature vs. nurture as a dichotomous choice. It is rather a question of degree. The biological constraints are real and should not be ignored. However, how strong these constraints are, and whether humans can overcome them if necessary, is another question. In the case of *Homo sapiens*, the evolutionary force of natural selection seems to have favored more and more "flexible programs" in the genetic materials. Oftentimes humans do not behave according to their instincts.[15]

What interesting and relevant hypotheses can sociobiology offer to our discussion? Interpretations and implications of sociobiology are often feared to be inherently racist.[16] Since sociobiology emphasizes natural selection and kin selection as the basic mechanisms for evolution, it highlights the present stratification of societies as the "natural order" or as the outcome of competitive forces with both winners and losers. Although evolution is opportunistic without any divine purpose, what we see is construed to be the result of the contest for survival of the "fittest" rather than the result of "cultural interactions." In this sense, sociobiology is a theory of status quo even if it offers no support for preserving the status quo in the future. Sociobiology is consistent with the postulate that, due to geographical separation and interbreeding, the forces of natural selection have produced a group of populations which are different in many important characteristics. The differences in physical and mental attributes may then be used in turn to explain (or to support) the differential positions that various groups occupy. Yet sociobiology does not really offer any insight or prediction as to precisely what these biological differences may be and may imply with respect to race.

Essentially the useful predictions on human behavior by sociobiology deal with the basic biological forces that the human species is subject to as a group. The sociobiological approac to behavior does not and cannot purport to explain subtle phen enological differences within the species. It is, therefore, a poor analytical tool in examining the question of race, particularly in a modern society. Sociobiology might be useful if we were looking at differences in human behavior which were essentially governed by biological forces. Specifically, we might see different behavioral patterns evolving among various groups when coping with different ecological environments. For example, we would expect that the differences in family organizational structures in different cultures could be explained to some extent by socio biology.

Perhaps sociobiology is most powerful in examining the question of sex,[17] where biology is indeed fundamental. Although technology is changing very rapidly, in this area humans yet have only limited control. Sociobiology can offer very color ful, if not always profound, explanations of male-female differences with respect to reproductive roles and related behaviors. (Since sociobiology assumes that the goal of life is to maximize reproductive fitness, the central focus is inevitably on reproduction.) Its key argument is based upon the difference between the two sexes in terms of "sperm-egg" asymmetry. Since each male has countless sperms and is physiologically capable of inseminating many females, he will attempt to maximize mating opportunities. The strategy for maximization leads to aggressive competition (Wilson, 1975:325). Only successful males survive to pass on their adaptive behavioral (genetic) traits to their offspring.

On the other hand, females have a limited number of eggs during their reproductive life. Thus each egg represents a large investment. In order to maximize female reproductive success, therefore, a good strategy is to maximize the female reproductive period by protecting herself without subjecting herself to risks and to wait for a "superior" male who would endow her offspring with genes which are more adaptive. In this way, her offspring have a better chance of surviving in the future. The male strategy involves risk-taking behaviors, whereas the female strategy involves risk-averting behaviors. The male strategy offers incentives to postpone puberty, whereas the female strategy offers incentives to hasten puberty. This is what we observe in humans as well as in other sexually reproductive species.

Sociobiology predicts that attitudes towards risks by the two sexes could be quite different, and this difference is based upon the biological asymmetry in reproductive functions and consequent strategies. Although this hypothesis is interesting and appears to be very useful with respect to animal behavior, it is not clear what constitutes "risk" in contemporary human society. Furthermore, it is difficult to separate biological factors from other cultural forces such as socialization and ideology. Nonetheless, casual observations may favor the hypothesis. Note the distribution of males and females in casinos, at slot machine areas, and elsewhere. It was not so long ago when men had to fight "dragons" to win the hearts of their ladies. No doubt socialization is a very strong factor (ladies are "not supposed" to do this and that), but one may argue that the role models are exactly what sociobiology suggests. Economists may find it interesting to speculate on the biological foundation of taste for risk, since they rarely examine it but assume as given. On the other hand, the implication of such differential attitudes towards risk by males and females is neither apparent nor unambiguous. It may be relevant if one is interested in modeling the behaviors between the sexes towards each other. For example, evolutionary theory predicts that in general human males have a tendency towards promiscuity, and females a tendency towards monogamy.[18] The theory may be useful for examining human behaviors of marriage, divorce, family size, sex ratio, adultery, a double standard, the generation gap, and the like. The theory predicts that the variance in male behaviors may be higher than that of females. Since males are willing to take chances, the expected return (payoff) from such undertakings may also be higher.[19] Does this mean that males will end up having more prestigious and higher-paying occupations as well as more powerful positions in society?

For one to justify these conclusions, it must be demonstrated first that the taste for risk cultivated among males due to their reproductive adaptive behavior does indeed apply to their attitudes towards risk in other economic (as well as sociopolitical) activities.[20] Second, it must be demonstrated that the different attitudes towards risk are due to something inherent in their nature and heritable rather than to the two groups acting rationally while facing different constraints. Expected returns from a risky activity are not the same for the two sexes.[21] Evolution may have selected males who are risk-takers and females who are risk-averters. Or evolution may have selected individuals who can analyze properly biological and ecological constraints in order to maximize their reproductive fitness. If that is the case, then we would expect increasingly similar behaviors and strategies taken by both males and females, as the genetically determined reproductive dimorphism becomes less important due to technological and cultural changes.[22]

In addition, it is extremely dangerous to focus on the mean tendency alone in discussing human behaviors in contemporary society, even if we accept that there is a biological force affecting males and females differently. After all, in many aspects there exists a tremendous diversity within each sex as is the case for race. Furthermore, the interactions of the biological and environmental factors (covariance between the two) may intensify the intragroup diversity. Although this remains to be validated empirically, we would expect that the overlap by the sexes is very large in most attributes and characteristics. The relevant question seems: Despite the biological forces at work equally on each group, why is there such a great diversity within the group, whether of sex or race, and what is its significance? After all, some women are stronger, more intelligent, and generally more able than some men.

The final point to be made is that the biological force no longer plays as important a role as it did in the past for *Homo sapiens*. We cannot ignore biology, of course, but we should not overemphasize it either. Human organisms have evolved and have become more complex than those of any other species. In the process, it is conjectured that the "programs" and "codes" genes carry have become more general in order to cope with more complex ecological and cultural environments. As a consequence, "brains" are taking over more and more "actual policy decisions, using tricks like learning and simulation in doing so" (Dawkins, 1976:64). The environment in which *Homo sapiens* lives is changing rapidly in terms of the ecological and, above all, in terms of the cultural setting. While genes do not learn, culture can. In short, culture is evolving at a faster rate than biology. In this sense the human species is caught between its biology and its culture, leading a "schizophrenic life." Barash (1977:324) speculates that some of our modern problems, such as alienation, pollution, overpopulation, intraspecific killing, and obesity, are symptoms of the "dissonance between the relative simplicity of our biological past and the diffuse complexity of our culturally mediated present." Our biology is not evolving fast enough to fit our cultural environment. In the framework of economics, this can be viewed as a problem of "principal-agent" relationship, where principals are the genes and the agents are human organisms. For human beings, the control exercised by the principal is diminishing rapidly in relation to the degree of free rein the agent has acquired.

Though biology remains an obvious force in human development, it may be no longer the determining one. In other words, biological determinism is becoming obsolete.[23] With the current changes in social attitude towards the two sexes and accompanying technology, we expect dramatic changes in traditional behavior patterns. To a great degree the biological force is under our control if we wish. Therefore, in many facets of life, sexual difference may now be considered irrelevant. However, in other areas of life, this is not the case, for, as Barash (1977:293) predicts, "We can afford equality of sexual behavior only when the biological consequences of such behavior are equal, but so long as women, not men, get pregnant, some differences can be predicted."

The dimorphism in procreation is crucial in determining equality not only in sexual behavior but also in social status. Reproduction, even with the help of modern technology, requires a large investment of time and energy. As long as females are expected to bear most of the burden of reproduction, and as long as the division of labor is carried out in the traditional way, the performance by the female sex in other fields cannot be the same as that of males.[24] Until the reproductive effort is recognized fully or shared by the other sex equally, females will be at a disadvantage. Since the unit of reproduction and social organization in our society is a family, equality of the sexes involves *intra*family reallocation of efforts and benefits. Of course, this process of intrafamily allocation can be made easier by institutional rearrangements at a societal level. To a degree, future generations are public goods. Institutionalized child care, a neutral or favorable tax system towards the working female, and maternity leaves for men as well as women are just a few examples of possible institutional rearrangements. More discussion on policy implications will be presented in the last section.

Since there is no biologically determined function assigned to each race, no analogy to sexual equality exists for racial equality. This is the first of two asymmetries emerging between race and sex differences.

A Simple Model of Distribution

In this section, we will discuss a simple model of distribution of economic goods in the post-Bakke world. In this world, we will assume that so-called "overt discrimination" based on race and/or sex is successfully eliminated in the realms of education, job hiring, promotion, and so on. In other words, we will assume equal treatment of the equals (in relevant qualifications) is in operation. However, this does not imply that covert discrimination is nonexistent.[25] Covert discrimination is difficult to identify and eliminate partly because we are not sure how our individual preferences are developed. When we have individuals who are different in any way, it is difficult to define what "equal treatment" of the unequals entails. This is the question of "equity" rather than of equality in the sense of identity.

As stated earlier, we would like to envision what distribution pattern may arise in the post-Bakke era in an economy where (1) meritocracy of some degree is accepted; (2) a market economy is the basic economic system; and (3) people do have concern for others.

In a static framework, wage determination in a particular labor market will be made by the forces of supply and demand. The demand for a particular type of labor is a derived demand which depends upon the productivity of the labor itself in production of goods and services and consumer demand for them. In turn, the marginal productivity of labor is a function of the amount of human capital embodied in labor as well as the amount of physical capital the laborer has available to work with. Therefore, it is very important to note that how productive one is depends upon both how much the individual has invested in himself, and how much the employer is willing to invest in production technology.[26]

On the other hand, the supply of labor to the labor market is determined by the individual's taste over work-leisure choice. In other words, how much the individual is willing to offer at each wage rate depends upon his taste vis-à-vis the consumption of other goods and leisure. If we assume, as economists usually do, that people work in order to earn income to purchase consumption goods, then the laborer is in effect buying his/her leisure time from the fixed initial endowment (i.e., maximum time available) in the Beckerian fashion. Thus, the supply of labor, given his/her preferences, is a function of the wage rate. However, the preferences are most likely "nonhomothetic";[27] then the supply of effort is no longer independent of the level of nonlabor income. In short, the initial level of wealth will influence the labor supply function. Extending this to the market supply of labor, we can conclude that the market labor supply is also dependent on wealth (both labor and nonlabor income) distribution. In addition, the decision as to how much to invest in human capital is also dependent upon the initial wealth. This is particularly true in view of the fact that the human capital market is imperfect, i.e., that the individuals may not be able to borrow funds to fully invest in themselves against their future income. The availability of scholarships and loans does reduce this gap; nonetheless, the quality and quantity of human capital investment will be greatly affected by the wealth as well as the income distribution of the people who finance the investment in human capital.

Now it is clear that both the supply and demand for labor

in a labor market are influenced by the initial wealth distributio·
This implies that equilibrium wage rates are also functions of
the initial distribution. If we assume that leisure is a normal
good, then high initial wealth will, due to the income effect,
reduce the supply of individual labor. This will in turn reduce
the market supply. At the same time, however, high wealth
will probably encourage human capital investment, and this
will increase the demand for labor. Thus, wage rates established
in the labor market may be a positive function of the average
level of wealth of the labor force in the market. At least in the
short run, where mobility of labor is limited, it appears that the
richer you are, the higher your labor income will be.

In the long run, this tendency will be counterbalanced by the
entry of new people into the market, increasing the supply of
labor. If perfect competition is assumed, competitive forces
from the new entry will lower the wage rate until the net wage
rate is equalized in all labor markets. Of course, practically speal
ing, the market may never reach this long-run equilibrium, since
mobility among occupations takes time and other factors may
shift the two curves in the meantime.

If there is any monopoly power being exercised by the labor
involved (such as closed shop union, or the license and degree
requirement), the zero profit (rent) long-run equilibrium of
perfect competition may never be established. A more reasonable model of factor (or any) markets is the one which moves
continuously from one disequilibrium to another. In facing
disequilibria, Schultz (1975:843) points out that "the ability
to deal successfully with economic disequilibria is enhanced by
education." In a rapidly changing economic environment we ma
say education (or human capital) is our way of adapting to the
environment. Evolutionary forces of natural selection may be
present here in a disguised form.

The discussion above now leads us to speculate on the pattern of economic distribution in the post-Bakke era. The main
point is that the distribution of income, occupations, and economic status is not free of the initial wealth distribution. This
implies that to the extent that the wealth distribution is lop-
sided among races in contemporary America, this unbalance
may be perpetuated, even if we successfully eliminate all forms

of overt discrimination based upon race and even if we are committed to providing some basic economic goods such as health, shelter, and education to everyone. In the post-Bakke era, therefore, the central issue involved in "economic justice" will be the question of wealth distribution. To the extent that some individuals are willing to work harder, happen to be in the right occupation, are more productive due to innate ability and/or education, and are willing to save a larger proportion of their current income, the resulting differential accumulation of wealth in one generation is inevitable. Hence, some degree of inequality is inevitable, and may be considered just, if one believes strongly in consumer sovereignty and meritocracy. However, this will certainly have differential impacts on the education of their children (e.g., private school vs. public school, university and/or graduate school), even if everyone is guaranteed a basic education regardless of parents' willingness to support it.

A yet more complex problem is how the inheritance of wealth to the next generations is treated. From our previous discussion, we must conclude that inheritance of wealth, if allowed, does lead, not only to intergenerational inequality but also to intragenerational inequality through its influence on the current income distribution.[28] In other words, this is a question of *inter*family distribution across and over time.

In order to deal with the question of "racial equality," we must therefore focus our attention on wealth distribution. Any social or economic policy for the promotion of racial equality is not complete if it ignores this aspect.

The analysis above, however, does not apply to the question of sex equality for obvious reasons. Since the basic unit of economic activities is usually a household,[29] both females and males do share the family wealth. In fact, it is not really possible to talk about the wealth distribution between men and women. It has been said that women are the only minority who live with the master race. The basic issue here is the *intra*family distribution of wealth in terms of human capital investment and labor supply decisions. In short, because of the very organizational structure of our society, the question of race and sex are not symmetrical with respect to wealth or property distribution.[30]

This is the second asymmetry between race and sex. Those who are involved in policymaking, therefore, should be aware of these two asymmetries.

Policy Implications and Conclusion

The previous discussion has been an attempt to contrast biological differences according to race and sex. It was demonstrated that biological differences in terms of race and sex do represent distinctive asymmetries in the areas of property distribution and reproductive functions in our society. In this concluding section, we shall examine policy implications derived from these asymmetries.

Minority races and women are said to have been at an economic disadvantages in the past. In correcting the past and present situation, however, it does not necessarily follow that the two groups should be dealt with in an equivalent manner. The asymmetries may imply and require different policies. Some of the possible policies will be discussed shortly, but first we shall discuss the issues of economic efficiency in allocation (of resources) and economic equality or equity in distribution (of economic goods).

In investigating the question of economic equality in the post-Bakke world, we have assumed that equity, meritocracy, as well as an approximately free market economy are valued as important. In such a society, it can be argued that some degree of economic inequality is tolerable—in occupations, income, and wealth—even if the inequality is due to differences in biological (i.e., genetic) differences of some sort. Differences in natural endowments such as physical and intellectual abilities as well as one's propensity for hard work (if it has any genetic basis) are probably justifiable grounds for a less than perfectly equal distribution of economic goods. Without differential rewards available, it is difficult to envision how an economic system is supposed to function efficiently based on incentives and market mechanisms.

In one sense, one may argue that "discrimination" based on race or sex may not be that different from selecting basketball players from taller people, jockeys from smaller people, football players from larger people, and assembly workers from people with more dexterous hands. On the other hand, it has been argued that discrimination based solely on race (or sex) is not justifiable in many situations since race (or sex) may not be a "bonafide" requirement for either the allocation or the distribution of economic goods in question but rather that the discrimination is an "insidious" form of interdicting a group of people from pursuing "happiness" on the basis of group affiliation.

The acceptance of the inevitably unequal outcomes arising from unequal distribution of biological endowments does not necessarily imply that any degree of inequality generated by the market mechanism is "just" or consistent with our society's notion of equity.[31] If the inequality is deemed to be excessive, the society may attempt to reduce the inequality by collective actions. How then do we reconcile this question of distribution with allocative efficiency?

In our meritocratic-market oriented but compassionate economic system, it is proposed that a clear distinction be made between the two functions of the system, i.e., allocation and redistribution. In the allocation, we may primarily depend upon the workings of market forces to determine who, what, and how. The "right" person at the "right" moment will receive high compensation for what she/he produces, whereas those who made the "wrong" decision (or were not so endowed to start with) will be compensated less for their efforts and achievements. The compensation is solely determined by one's merits in the marketplace regardless of need. To the extent that one's merits are under his/her control (such as the amount of effort exerted), one may deserve the "fruits of industry (productivity)." As long as we assume that people are responsive to rewards, we run the risk of losing incentives without meritocracy.

Consequently, it follows that an individual with many dependents is no more entitled to better jobs or higher salaries than those without dependents solely on the basis of the "needs." At the time of hiring and promotion, for instance, the wife of a wealthy man should be judged solely on her merit as a worker, and not by her need or lack of need for status, income, and the like. Otherwise, it is extremely difficult to determine whether or not an individual is compensated "fairly," whether or not equal pay for equal work exists, or whether or not equal opportunities exist for everybody.

On the contrary, we may wish to redistribute total products and wealth among ourselves in order to create a more "equitable" system.[32] If the distributive outcome of the market mechanism is deemed to be inconsistent with social goals, then proper adjustments and corrections are required. Given market imperfections, uncertainty, stochastic elements in life, and other factors, there is very little reason to assume that the markets alone will achieve even the "second best" situation. Redistribution would serve two purposes: first, to correct, if necessary, the intragenerational distribution (market outcome); and second, to correct the intergenerational distribution.

Of course, our familiar efficiency-vs.-equity problem arises at this point (Okun, 1975). To a certain degree, efficiency and equity are not compatible, hence a trade-off problem. How efficiency is influenced by incentives is an empirical question, since the degree of the trade-off is dictated by the preferences of the population under consideration. If the underlying preferences are such that wage rates do not enter into the work-leisure decision, it is not necessary to worry about the incentive question. Of course, this implies a rather peculiar type of preference, upon which most economists would be hesitant to rely.[33]

In the post-Bakke world, discussion of the mechanisms for allocation and redistribution for "racial equality" (as opposed to sex equality) is relatively straightforward. Once the game is deemed more or less fair because the players have approximately equal opportunities to prepare and to participate, the resulting outcomes may also be considered fair. This raises a serious question as to what a "fair" starting point involves and what equal opportunities entail. Among other things, the questions of interracial wealth distribution and inheritances must be resolved. Once done, the biological differences due to race will become irrelevant, and the differential outcomes are solely the result of individual differences. And these individual differences we will have forever. If evolutionary theory tells us anything, it is that diversity is the key to the species' survival.

Redistribution may be accomplished mainly by tax reforms such as provisions of tax incentives in favor of minority races, very progressive income taxes, and very severe inheritance taxes. In addition, many policies to strengthen basic education (such as equalization of school funds, special programs for minority races, busing and other schemes for integration[34]) and higher education (such as scholarships, low interest loans, and affirmative actions) may be instrumental in raising the level of human capital of the minorities. However, all efforts towards cultural integration in the spirit of the "melting pot" may have a costly side effect in the loss of ethnic culture.

The more inherently complicated question arises when one looks at the problem of sex. Here the central issue is that of reproduction, i.e., intrafamily redistribution of resources. To the extent that women and not men are incapacitated by pregnancy, birth, and lactation, their job performance will be compromised. The question is: Is this a personal problem of the individual woman, a problem for the couple involved, or a problem for the general society? If the woman is solely responsible, she will be at a disadvantage in the marketplace. We will expect unequal representation by and remuneration for the woman. This is probably not a very popular idea, since women cannot reproduce alone.

If the parents are equally responsible for their offspring, then the intrafamily redistribution of responsibility may be required for sex equality. The current economic system is biased towards the traditional division of labor, not to mention our traditional expectations and role models. However, it does not require too much change to neutralize the institutional constraints. For example, the burden of childbirth and rearing can be alleviated significantly for mothers if maternity leaves for men as well as women become customary.[35] The present tax bias in favor of women (or men, for that matter) staying at home can be modified. Such institutional changes should make it easier for men to stay home if they wish. If two working parents is to be a norm, then the reduction of working hours for each or alternatively flexible hours will ease the burden tremendously. Better day care, nurseries, and after-school care are obviously essential.

If offspring are considered to be a public responsibility, then society at large can engage in activities which assure both men and women of equal opportunities. Publicly subsidized day care, nurseries, and after-school care are examples. In addition, collective action may be taken for baby-sitting services, cleaning services, and so forth. These services may be made available free, on a subsidized, or user-cost basis.[36]

If reproduction is to be carried out with a more equal division of labor, then this will probably entail more equal division of responsibilities in quasi-reproductive and nonreproductive areas such as child-rearing practices, custody rights, divorce and alimony settlements, community property arrangements, and social security benefits. The proposed constitutional amendment embodied in the Equal Rights Amendment is certainly a step toward this direction.

It is hoped that possible policies mentioned above are sufficiently suggestive to point out the crucial asymmetries underlying in the economic status of minority races and women. Race and sex are indeed both biologically and socioeconomically significant in our society, and have been important factors in creating and preserving economic inequality among different groups. In the case of racial differences, due to historical and institutional development, the minority races, particularly the blacks, have been forced to a lower economic status in our society. All forms of explicit discrimination are now declared to be illegal. This paper has argued, however, that legal changes in the forms of civil rights and antidiscrimination are necessary but not sufficient to eliminate interracial inequality, since property distribution (as the result of the differential accumulation of capital in the past) is already skewed enough among races to negate any possibility for "fair competition."

Contrary to racial differences whose significances are more sociopolitical than biological, sex difference implies an immutable biological difference in reproductive functions. This biological difference is potentially more difficult to reconcile in our society, since it is not easy to obtain consensus among the members of this society as to economic roles assigned according to sex. The position taken in this paper is that if we want to have more social equality between the sexes, particularly in the marketplace, then the reproductive responsibility must be shared by both sexes more equally than in the past.

In this way both men and women will be able to experience equal rights and equal opportunities in search of "life, liberty, and the pursuit of happiness." Implicit within these rights will be each person's freedom to choose his/her occupations from among many different careers. Indeed, to choose to be "houseperson" may, in that sense, be seen as an exercise of these most fundamental human rights.

Notes

1. See Bird (1971:1-15) for an interesting account of how "sex" was introduced.
2. The reproductive isolation between members from two groups establishes two distinct "species."
3. According to Goldsby, this can be in the order of 80^{23}, if we assume that there is a 10 percent variation in genes and that each person has 10,000 genes.
4. It is conceivable that stigmatization of a group may occur if that blood type is identified as socially undesirable or inferior.
5. "True hermaphrodites" are exceptions. Of course, taxonomy may be somewhat arbitrary.
6. Lillard (1977) reports that the level of inequality in lifetime human wealth is much smaller than that in current earnings or wealth.
7. One may follow the classification scheme by Goldsby (first chapter in this book).
8. A somewhat disturbing piece of statistics is that the average IQ score of blacks in this country has been about 15 points below that of whites in the last sixty years (1918-1971). At first glance, this difference appears to be surprisingly large and consistent. However, one needs to be very cautious in interpreting this number. A bias in the testing method is possible (Welch, 1976). There are three additional questions to be answered: What does it measure? How stable is the score? How is it relevant to economic equality?

It is well known that the IQ test is designed to be and is a reliable predictor of the subject's level of school performance. (At the beginning, it was designed to predict whether or not a student will finish school.) Whether it measures "general intelligence" or "innate ability" is still highly controversial.

Scarr and Weinberg (1976) report that the black children adopted and raised by white families score five to ten points higher than their counterpart raised in black families. Schratz (1978) found that, given tests on spatial ability and mathematics, preadolescent subjects scored quite differently. The mean scores for both Hispanic and black women were higher than those for both Hispanic and black men, whereas the mean score for white men was higher than that for white women.

Since schooling is a very important determinant of productivity and is used as a screening device, the IQ score is also expected to be a powerful predictor of earnings. However, in view of findings such as the above, one could not give much credence to IQ as the indicator of natural ability of different races.

9. Blumberg (1977) has shown that males and females respond differently to hepatitis virus, for example.

10. For a critical examination of sociobiology, see a collection of essays in Montagu (1980). Chagnon and Irons (1979) and Lockard (1980) are excellent sources for the analyses of variations in human behavior from anthropological points of view.

11. The difference may not be as great as it may seem, for economics does deal with the welfare of future generations. Bequests are common. Environmentalists are presumably concerned with the welfare of future generations. It may be a matter of a difference in emphasis. In addition, we often observe behavior which is consistent with the predictions in sociobiology that individuals are more concerned with the welfare of individuals who are to some degree genetically related. The altruistic behavior may decline progressively as the degree of relatedness diminishes. See Hamilton (1964) for Hamilton's rule.

12. "Pleasure-pain" calculus was a major concern by the nineteenth-century moral philosophers and economists such as Bentham, James Mill, and J. S. Mill. The term "pleasure" could be an inclusive term involving not only hedonistic but also broader psychological and intellectual pleasures in addition to more fundamental and "utilitarian" concepts.

13. The theory of kin selection, on the other hand, may be used to explain these behaviors in a fairly convincing manner.

14. "If biology seems arrogant in seeking to include humans within its scope, think of the greater arrogance of a social science that refuses help when it is offered" (Barash, 1977:277). "Sociobiology has achieved instant popularity in part because the better-known social science research strategies cannot provide scientific causal solutions for the perennial puzzles surrounding phenomena such as warfare, sexism, stratification, and cultural life-styles" (Harris, 1980:333-34).

15. Barash defends sociobiology to be neutral. However, the interpretations can be abused to support "racist" evaluation of different races.

16. Or rather our instincts are not sufficiently keen to guide us even in our daily life. For example, we don't know what is edible and potable unless told.

17. Sociobiology is "sexist" in that it recognizes male-female differences

which are "genetically coded" and not learned (Barash, 1977:283). For more discussion on human sexuality through sociobiology, see Symons (1980) and its critical review by Geertz (1980) as well as part IV of Chagnon and Irons (1979).

18. "Which of these two tendencies wins in particular societies depends on details of cultural circumstances, just as in different animal species it depends on ecological details" (Dawkins, 1976:177).

19. Reflecting on human evolutionary history, Symons (1980:176) speculates that evolutionary force promoted "substantial differences among males in competitive abilities" in societies where hunting was a primary economic activity of the adult males. The rewards for successful hunters were often women who in turn often represented a sign of status.

20. Taste for risk is considered important by economists as a determinant of occupational choice, income distribution, etc. Risk-lovers will choose opportunities which offer a small chance of high returns whether it is fame, prestige, power, or wealth. A few will succeed and the rest will lose out. The returns to the group of risk-lovers thus is expected to range widely, i.e., a large variance. If most people are risk-averters, then risky jobs will have higher returns to compensate for the risk. For more details, see Weiss (1972), Tobin (1958), Von Neumann and Morgenstern (1953) among others.

21. This is no doubt a very powerful explanation for the unequal distribution of occupations, political offices, and self-made millionaires between the sexes in the past.

22. Anthropologists have observed that male dominance is almost universal both in contemporary and ancient societies. For a discussion of how reproductively related division of labor leads to subordination, see Rosaldo and Lamphere (1974), particularly the first three essays.

23. One of the important contributions made by sociobiology is to direct our attention (i.e., social scientists) to the importance of genetic influence in behavior. Many social scientists have been accused of cultural determinism.

24. Van den Berghe (1980:83) claims that "There are good biological reasons why women should raise children more than men, but there are no good reasons why women should not be airline pilots, or men typists." Thus the occupational distribution may be independent of sex. However, the difference in energy available and devoted in occupational development will certainly affect the performance and hence the payoffs.

25. Covert discrimination or secondary discrimination is much more difficult to define and substantiate. It arises in the socialization process, development of role models, and so forth.

26. Unless the productivity of labor is independent of the amount of capital.

27. Nonhomothetic preference implies that the trade-off an individual

is willing to make over work-leisure choice depends upon the level of income.

28. Brittain (1977) found a high correlation between the parents' socioeconomic backgrounds (including material wealth and education levels) and those of the offspring. In addition, there was considerable evidence of assortative mating in his sample.

29. Of course, a household can consist of a single female. The evolutionary theory supports monogamy for species which need large investment in child rearing by both sexes. Reproductive activity may require not only energy but also time. In order to obtain self-sufficient offspring, the female-male team may spend a large portion of their reproductive life together.

30. Of course, there are families which are racially mixed although they are classified somehow into one racial group by the census interviewer.

31. Of course, it is not always easy to achieve consensus.

32. Redistribution may be made in order to provide (1) the minimum amount of public goods or (2) the "primary goods," by Rawls (1971), such as rights, liberties, powers, opportunities, self-respect, income, and wealth. In the latter case, unequal distribution of any, or all, of these values is admissable only if it is deemed to be to everyone's advantage.

33. Although people are not always able to choose their hours or even occupations due to institutional constraints.

34. Whether busing and integration schemes are the most effective methods of strengthening basic education is a subject for empirical investigations.

35. Presumably without compensation if the individual family is to finance reproduction.

36. Charlotte Gilman (1966:242), who was very influential in the women's movement around the turn of the century, suggested in her "Women and Economics," published in 1898, that if a "commodius and well-served apartment house for professional women with families" were built, "it would be filled at once." These apartments would be without kitchens, and meals would be professionally prepared by the household industry either catered to the homes or served in the common dining room. Similarly, other household chores are done by professionals.

7

Social Equality and
Ethnic Identity:
A Case of Greek-Americans

SANDRA L. SCHULTZ

Introduction

The notion of "social equality" within a pluralistic society involves as a necessary concomitant the concept of equal opportunity for all individuals regardless of ethnic affiliation. The ideal of social equality is expressed in the familiar "all men are created." The real situation is, however, that there is inequality in the United States, inequality in regard to employment, housing, education, and treatment by legal and governmental institutions. Often, inequality stems from a group's ascribed characteristics, rather than an individual's achieved characteristics. This essay will examine the interplay of ethnicity and social equality. Specifically, it will examine the ethnic identity of one group of Americans, Greek-Americans.

The term "ethnic identity" follows from the notion of "ethnic group."[1] That ethnic groups provide their members the means to "identify themselves in the face of the threatening chaos of a large and impersonal society" has been expressed by some (Greeley, 1971:52). "Identity" is often used synonymously with belonging, membership, social attachment, and group affiliation. Similar to the concept of racial groups, the concept of ethnic groups is often based upon biological (such as skin color) as well as cultural characteristics (religion, national origin, etc.). It is sometimes difficult to separate these components.

Thus, to consider the interplay between biological difference and social equality is necessarily to consider the interplay between cultural (or subcultural) difference and social equality.

Although members in an ethnic group are tied by some factors in common, the group may be culturally heterogenous. The dominant societal group, however, often ignores cultural variation within ethnic groups. In the United States, social policies generally operate on two assumptions: (1) that differences within ethnic groups do not exist, and (2) that differences between ethnic groups should be ignored. In short, these policies ignore (so deny) social differences between and within ethnic groups. It is assumed by policy makers that biological relationship (of varying degrees) implies social similarity.

This essay takes issue with the second assumption and suggest that a perceived ethnic group member may not be representative of the ethnic groups. The assumption of ethnic group homogeneity is examined. The specific case of Greek-American ethnicity is presented in order to illustrate the operation of ethnic group heterogeneity.[2] It will be shown that, for this example, ethnic group members recognize important social differences within their group. Essentially, the issue raised here is will the policy of ignoring intraethnic group differences bring about the goal of social equality for individuals? Various existing schemes seeking to achieve social equality may be insufficient to achieve their stated goals.

In the next section, Greek-American self-identification will be presented both in terms of similarities and of differences within the ethnic group. The purpose of this section is to provide a sketch of representative views held by Greek-Americans who have been interviewed extensively. The last section discusses policy implication of these similarities and differences and then presents conclusions.

The Greek-American Study Community

This section on Greek-American ethnic identity is based upon the information collected in 1975 as part of a larger ethnographic project. The primary purpose of the project was to investigate the interrelationship between ethnic identity and

marriage. The study population consisted of a community of
about 150 Greek-American families. The community is a part
of a southeastern city in Arizona with a population of 400,000.[3]

The word "community" in this investigation (and in all
other works dealing with Greeks in the United States) derives
from the Greek concept *kinotis*, meaning an organized colony
which centers around a Greek Orthodox church. The Greek-
American community is frequently incorporated under state
law. Membership is open to all Greeks residing in the city or
district where the community exists. Such communities are
essentially autonomous bodies, functioning independently and
without obligation to a central Greek Orthodox archdiocese.

The following picture of Greek-American ethnic identification
is the product of forty-six formal interviews as well as of many
informal conversations with Greek-Americans. In these interviews, informants were asked to characterize a "Greek" (or
"Greek-American"; the two are used synonymously to apply
to anyone of Greek descent). Though not every informant
mentioned every feature listed below, there was surprising agreement on the attributes of Greekness. Those interviewed were
not led by the interviewer in their appraisals but were asked
open-ended questions such as, "How would you describe a
Greek man?" In addition, descriptions of informants' experiences
with other Greeks and Greek-American communities were noted.
For instance, in a conversation unrelated to the task of describing "a Greek man," an informant commented that a wealthy
Greek friend of his was too proud to admit his upbringing in a
coal-mining town. This comment was noted, and after a series
of many conversations in which Greek honor was mentioned,
it became obvious that honor was a feature crucial to Greeks'
(Greek-Americans') perceptions of themselves.

In what follows, an attempt was made to provide a synthesis
of the large amount of information collected through the interviews and conversations. Much of the information is qualitative.
Rather than presenting a detailed report of the raw data and
their statistical attributes, various similarities and differences
are highlighted as an illustration of one ethnic group. It would
be clear that many of the characteristics discussed here in the
context of Greek-Americans in a particular community do apply,

to a large extent, to Greek-Americans in other communities, and possibly to other (non-Greek) ethnic groups.

Intragroup Similarities

The attributes which Greek-Americans see as characteristic of themselves (*common* to group members) are presented in this section. They will be discussed as "our heritage" (tradition, Europeanness, religion), "our people" (kinship, cliques, stereotypes of Greek men and women), and "our nature" (honor, emotionalism, envy, jealousy, fear, and distrust).

One aspect of Greek-American identity can be labeled "our heritage," under which the dimensions of tradition, Europeanness, and religion are subsumed. Greek-Americans in our sample make this distinction between themselves and others. They recognize both the distinctiveness of other ethnic groups in the United States (such as "Serbians") and the cultural distinctiveness of groups defined as citizens of nations (such as "Italians") Most citizens of the United States are classified by Greek-Americans as "Americans." "Americans," a label applied to a category in opposition to "Greeks," connotes a people without tradition, without emotion, without flair. Greeks, on the other hand, do have tradition and emotion. In short, they have a distinctive zest and gusto. In comparison, "Americans" are thought to be extremely dull.

Greeks in the United States tend to think of themselves as culturally distinct from Americans. They see themselves as being culturally "European," rather than culturally "American." For instance, it is explained that Greeks, being culturally European, customarily shake hands more often than do Americans. Greeks, being culturally European, are more protective of women and girls than are Americans. Greeks in America (the first and second generations at least) more closely identify with Europeans than with Americans.

Another element of Greek heritage which distinguishes Greek from most Americans is their Eastern Orthodox religion. This also sets Greeks apart from most other Europeans. Thus, Eastern Orthodoxy, and specifically Greek Orthodoxy, contributes to Greek ethnic identification. Many Greek-Americans do not conceptually separate Greek culture from Greek Orthodoxy. "To find one's heritage" was equated by an informant with sudden, active participation in Greek Orthodox religious ritual. The Greek Orthodox church in the United States is recognized as a vehicle of Greekness, an institution that promotes and maintains Greek identity. Aside from its religious function, the church is viewed as a meeting place for Greeks.[4]

Closely tied to Greekness and Greek religion is "tradition." Tradition is conceived of as a constellation of rules governing responses to certain interactional situations (encounters with mortals and with the supernaturals). Such situations are repetitious; they have happened in the past and will happen again in the future. Tradition equips individuals with time-honored rules of comportment so that they may correctly respond to recurring situations. For instance, observations of religious holidays is dictated by tradition. Courtship and marriage practices are said to be governed by tradition. It is said that Greek-Americans are "keepers of tradition," that having a sense of duty and tradition is characteristic of Greeks. Those Greek-Americans who "do not care for tradition," who do not exercise Greek traditions, are spoken of as "Americanized."

A second aspect of Greek-American identity can be labeled "our people." The following topics are included here: kinship, cliques, and stereotypes of Greek men and women.

The importance of family ties (kinship) pervades the story of Greek migration to the United States. Greek men and boys journeyed to the United States as husbands and sons in order to earn money to contribute to family coffers. By and large, the first wave of Greek immigrants to the United States was comprised of Greek males, mobile wage earners who contributed much of their earnings to their families in the homeland (Fairchild, 1911; Saloutos, 1964). Sometimes, these early Greek pioneers sent for their families. This practice is paralleled today by recent Greek immigrants to the United States. Also, migration patterns of Greeks within the United States seem to continue to follow the old form of chain migration of the family unit.

The Greek seems to identify himself and seems to be identified by other Greeks in relation to his family. It is the family

that receives praise, not the individual. A common expression
is "That family has distinguished itself."

Closely related to the notion of family is that of "clique,"
also called "clan." A clique is a group of people who congregate
for entertainment, who acknowledge a common bond, and
who display loyalty to each other. Though kinship undoubtedly
is a primary bonding agent, other social factors function to
produce clique bonds. Regional affiliation in the homeland (for
instance, ancestry in Crete) can be important in bringing people
together (Theodoratus, 1971). It is said that Greeks are "clannish"

Informants have clear pictures of "the Greek woman" and
"the Greek man." The Greek woman as a proper Greek wife
"stays home, raises the children, and shows her bracelets." She
has been described as quiet, pretty, submissive, religious, afraid,
and subdued. The Greek woman is expected to assume the roles
of subservient wife, good mother, good housekeeper, and proper family representative. Informants also hold a clear picture
of "the Greek man." He has been described as stubborn, chau-
vinistic, selfish, proud, dominant, loud, authoritarian, lazy, and
emotional. The Greek man is expected to assume the roles of
domineering husband, family provider, and opinionated, temper-
amental man of the world.[5]

Another aspect of Greek-American identity can be labeled
"our nature," under which the dimensions of honor, individualism, emotionalism, envy, jealousy, fear, and distrust are subsume

"Honor" (the important Greek concept *philotimo*) refers to
self-esteem and may be manifested on the level of the family
or household. A person entering a Greek home is honored, and,
in turn, the family is honored (or preserves its honor). Because
of the necessity of maintaining family honor, Greeks do not
air family problems publicly, because to expose family problems
is to bring shame upon the family and upon each individual
family member (Campbell, 1965; Du Boulay, 1974; Friedl,
1962; Lee, 1959).

Possession of family honor has been spoken of as possession
of pride. Greeks speak of themselves as "a proud race." Greek
pride operates on many levels: the individual, the family, the
ethnic group, the homeland, the heritage. Persons of Greek
ancestry are said to be "proud individuals" and emotional.

Individualism and emotionalism are responses which protect honor. The self-identified characteristics of envy, jealousy, fear, and distrust are also manifested in response to maintenance of honor. (See Stycos, 1948, for comparative information on New England Greek-Americans.)

The fourth self-identified characteristic of Greek-Americans is possession of intelligence. It is said that Greeks are intelligent by nature and that the intelligence of Greek immigrants has been sharpened in the United States by hard work.[6] The image of Greek-Americans as hard workers is sustained by many rags-to-riches stories; each of the Greek-Americans formally interviewed could tell at least one dramatic success story about fellow Greek-Americans. Intelligent Greeks, it is believed, through hard work achieve such valued accomplishments as education and wealth. Greek-Americans view themselves as "money-conscious."

In summary, our data suggests that Greek-Americans possess a strong ethnic identity. This identity makes claims about group origins ("our heritage") and about behavior of group members ("our people") and expresses contrasts with other groups ("our nature," "our intelligence"). From the discussion thus far, one might infer that Greek-Americans see themselves as a homogeneous group. However, the aspects of ethnic identity presented so far have told only part of the story. In the following section, self-identification of intragroup differences is presented.

Intragroup Differences

Neither Greeks in the United States nor Greeks in the homeland are homogeneous groups. Informants recognized this and readily discussed differences between Greek-Americans and Greek nationals. (The first set of self-identified differences is relevant to this essay.) Our Greek-Americans while describing similarities among themselves nevertheless perceived differences among themselves. In short, they perceived a heterogeneous ethnic group. The following differences between Greek-Americans, as seen by Greek-Americans, will be discussed: generational differences, regional differences, class differences, and differences between "oldtimers" (and their descendants) and "newcomers" (and their descendants).

The generational time depth of Greeks in the United States is three; there are first-, second-, and third-generation Greek-Americans. (Greeks immigrating to the United States after the Second World War form a new and contrasting first generation. These later arrivals are classified separately from earlier arrivals and will be discussed below.)

Results of assimilation were discussed by Greek-Americans, where the first generation was said to be "more Greek" than subsequent generations. It is believed that quantity of Greekness is progressively diluted in each generation born away from the homeland, that generations lose Greek characteristics in adjustment to the host society. The first generation is said to cling to Greek culture and Greek Orthodoxy. Average members of the first generation are described as superstitious, emotional, and not well educated.

Second-generation Greek-Americans, like their parents, tend to have strong attachments to Greece, but they "consider Ameri ca first." It is said that the second-generation Greek-Americans are extremely money-conscious, educated, and, in some cases, choose not to be part of Greek-American communities. Discussion of second-generation Greek-Americans concerned loss of Greek characteristics as well as gain of non-Greek characteristics The latter process was illustrated by comments that some second-generation parents institute new patterns of child-rearing

The third generation of Greek-Americans is considered to be more changed culturally than the second generation. Members of the third generation are said to be "assimilated" or "thoroughly American." On the other hand, it was acknowledged that the third generation does possess Greekness and that it would lose much of this quality were it not for formally organized Greek-American communities. In other words, while members of the third generation are "less Greek" than their grandparents, because of the existence of Greek-American communities, they are "more Greek" than they are "American," "Serbian," "Italian," and so forth.[8]

Greek-American informants recognize regional differences existing between people living in Greece and in former Hellenic territories. Differences in dialects, foods, surnames, and styles of dancing and singing are noted. To some informants, regional differences may also involve temperament, morality, and concepts of ethnic purity. Greek immigrants are believed to carry with them regional traits as they do traits of Greekness. Regional affiliation, therefore, is an important dimension of an individual's identity. Regional antagonisms are transferred to the host country, as are practices of regional stereotyping. For instance, Greeks from Asia Minor are held in contempt by some Greeks from the Peloponnesus. Their long association with Turks (traditional enemies of Greeks) and their rural ways provide the rationale for this. In comparison, the Greeks of the Peloponnesus have been less influenced by Turkish culture (so they claim) and believe themselves to be more purely Hellene. Aegean island Greeks are likely to regard Peloponnesus Greeks as backwoods bumpkins.[9]

Another kind of regional difference should be considered here: regional differences which have developed within the United States. Americans in general recognize regional differences (southerners, westerners, and so forth), though American regional identification may be diminishing through operation of several factors. It is likely that Greek-American regional identification may also be diminishing. Greek-Americans have been in the United States only a relatively short time. Theirs is a strong ethnic identification which may supersede the adopted regional component. There are strong Greek-American secular and religious associations which stress Greek-American identity. Finally, the influence of mass media and industry on American society in general may have tended to weaken existing regional boundaries and characteristics.

Greek-Americans recognize two major socioeconomic classes which have existed and continue to exist in Greece. These are the upper class (associated with education and with urban residence).[10] To a certain extent, homeland class affiliation continues in the first-generation Greek-Americans. The label "the R. Group" (meaning "the Rural Group") is used by some second-generation Greek-Americans to refer to uneducated, conservative Greek immigrants. One can be designated as a member of "the R. Group" regardless of the length of time one has lived in the

United States. In essence, the label denotes membership in the Greek lower class. American-born generations can outgrow the stigma of their ancestors' rural Greek origins through education Consequently, the label "the R. Group" is seldom applied to second-generation Greek-Americans.

This is not to say that class differences are absent among second- and third-generation Greek-Americans. Class differences are present. The criteria are wealth and education. It should be recalled from the previous section that Greek-Americans view themselves as "money-conscious" and see education as a means of upward mobility. Education, its presence or absence, is a criterion for judging individuals and groups (families, Greek American communities). Educational attainment is also, in Greek-American eyes, a measure for evaluating the quality, progress, or progressiveness of ethnic groups.

Perhaps the intragroup difference most vividly expressed by Greek-Americans is the difference between early Greek immigrants (and their descendants).[11] This distinction was articulate by the descendants of the early (1880-1924) Greek immigrants; their images of Greeks immigrating to the United States after the Second World War are anything but complimentary. The postwar immigrants (called "DPs," meaning "displaced persons" are described as unappreciative of older immigrants, envious of the wealth and education ("progress") of the "oldtimers," selfish, lazy, and lacking humility. It is said, "The newcomers came and found the table set. They ate and then said the food wasn't any good." Though both sets of immigrants are technically "first generation," the descendants of the early immigrants made no bones about the difference between the two groups. The newcomers are said to be a "different breed" entirely from the oldtimers. Whereas the oldtimers were uneducated villagers who lacked the benefits of formal education, they (being clever Greeks) overcame many disadvantages and became successful.[12] They were hard workers. They provided for the education of their children and sent money home to Greece to build sisters' dowries and brothers' educations. Newcomers, on the other hand, are viewed as freeloaders who act as though the oldtimers owe them something. It is said that they refuse to work hard and are content to "make a comfortable dollar today, and they don't worry about tomorrow."

This view of the newcomers reflects another perceived difference: the difference between Greeks in the homeland and Greek-Americans. Essentially, achievement through hard work divides the two. While all Greeks are thought to share certain dimensions of identity ("our heritage," "our people," "our nature," "our intelligence"), the dimensions of hard work (and, thus, achievement) pertains primarily to Greek-Americans. Specifically, it applies to oldtimers. While some newcomers are recognized as hard workers, most newcomers (as well as Greeks in Greece) are thought to be lazy. They have not pursued the ethic of hard work; they have not made the most of their potential; and they have not achieved (wealth, education, high social status) as have the oldtimers.

The offspring of oldtimers regarded newcomers as conservative in religious practices and slow to integrate into American society (or Greek-American society). Some said that their Greek community was "fifty years behind" other communities because of the large percentage of postwar immigrants. The newcomers responded that the second-generation Greek-Americans (offspring of the oldtimers) were "Americanized," formed cliques, and were generally inhospitable.

The perceived differences among Greek-Americans has been discussed in this section. Neither Greeks in the United States nor Greeks in the homeland can be said to be homogeneous populations. Greek-Americans recognize this and incorporate the concept of heterogeneity into their ethnic identity.

In summary, Greek-Americans recognize generational differences among themselves, where the first-generation Greeks (the immigrants) are thought to be "more Greek" than subsequent generations. It is believed that the third generation would lose much of its Greekness if it were not for the presence of formally organized Greek communities in the United States.

Within Greece, regional differences are emphasized. Regional differences are said to involve language, custom, and morality. Beliefs about regional differences are operative to some degree in the United States and act to divide the ethnic group.

Class differences are recognized as existing in the Greek populations of both the United States and Greece. In Greece, upper-class status is determined by wealth, education, and urban residence, whereas lower-class status is determined by lack of wealth, lack of education, and rural residence. The first American-born generation can "outgrow" the stigma of parental lower class affiliation, but immigrants from rural Greece cannot shed the label "the R. Group."

Differences between early Greek immigrants and later Greek immigrants (immigrants after the Second World War) are claimed by the descendants of the oldtimers. They criticize the attitu and behavior of the newcomers who "ate and said the food wasn't any good." The clash between the descendants of the oldtimers and the newcomers arises over the ethic of hard work Though the newcomers behave in accordance with rural Greek culture, calling upon bonds of kinship for economic aid, second generation Greek-Americans interpret this behavior as taking financial hospitality between kinsmen for granted. The oldtime offspring resent the demands of the newcomers and their criticisms of life in the United States.

Discussion and Conclusions

The Greek-American case has been presented above as one example of ethnic identity incorporating notions of intragroup similarity and difference. Ethnic identity expresses beliefs abou group origins and about behavior of group members. The latter expresses rules of comportment as well as contrasts between the behavior of members and nonmembers. Thus, similarities between group members are described. For many large ethnic groups, an equally important component of identity is recognition of intragroup differences. Such factors as complexity of the homeland society, complexity of the adopted society, processes of culture change in home and host society, coupled with continued immigration, operate to produce a heterogeneo ethnic group. Heterogeneity is often expressed in terms of generational, regional, and socioeconomic differences. It would seem that the heterogeneous ethnic group may well be the rule rather than the exception.

This picture of heterogeneous ethnic groups has implications for social policies devised to achieve social justice in various arenas of American society. Most such social policies are blind to differences within ethnic groups, that is, they assume the homogeneity of ethnic groups and operate under the principle that any ethnic group member is equivalent to any other ethnic group member for policy purposes.[13] The Greek-American example demonstrates that this may not be the case. The policy-making outsider (nonethnic), lacking sufficient understanding of the possibility of ethnic group social structure and heterogeneity, forces his classification of "minorities" upon ethnic group members. In the Greek-American case, an educated and wealthy third-generation Greek-American whose ancestors came from Crete is not culturally equivalent to a relatively uneducated first-generation Greek-American, born in Epirus, who arrived in the United States as a war refuge. Policies which do not differentiate subdivisions within ethnic groups are not based on social reality.

In some cases, subethnic divisions may be more crucial to the problem of social injustice than gross ethnic divisions. For instance, a first-generation Greek-American may be discriminated against more than a second-generation Greek-American. Or a southern Greek-American may be discriminated against more than a northern Greek-American. The implications for other ethnic groups are obvious.

Given the likelihood of ethnic group heterogeneity, those who devise social policies (e.g., equal-opportunity policies) can take either of two directions. They can accept the possible inability of their present schemes to achieve social equality for individuals and be content with the belief that social equality is being attempted on the level of the ethnic group. Rejecting this, they can evaluate social programs by asking whether these programs are reflective of social reality. Must an individual clumsily fit himself into the outsider's ethnic group classification scheme? That is, are the criteria of group membership imposed from the outside or generated from the viewpoints of insiders?

Social policy is defined as "planning for social externalities, for the redistribution of social benefits" (Rein, 1970:5). Makers of social policy are reliant upon the input of political scientists, economists, sociologists, anthropologists, in short, social scientists, to guide them in decision making. Sociologists and anthropologists in particular can provide valuable information on social stratification, component social groups, and the relationships between component groups. Sociologist Nathan Glazer (1975:31) has said, "To redress inequalities means, first of all, to define them. It means the recording of ethnic identities, the setting of boundaries separating 'affected' groups, arguments among the as yet 'unaffected' whether they too do not have claims to be considered 'affected.'" Research is needed on the dynamics of group definition, on majority and minority relations. For instance, how do different types of minority groups define and approach the societies around them (Newman, 1973)? In what way are social meanings attached to perceived differences between groups? Dale L. Hiestand (1970), who has studied the issue of discrimination in employment, has expressed the need for research in the following areas: (1) on the significance, in terms of competition, of two or more big minority groups in the labor market; (2) on the distinctions faced by different socioeconomic sectors of a single ethnic group; (3) on the difference between established urban ethnic families and new (rural) immigrants; (4) on the comparison between different minority groups with respect to their relationships to potential customers; and (5) on the differences in employment between minority middle-class and majority middle-class members. In addition, little is known about the "white ethnic groups" in the United States (Glazer, 1975; Levine, 1971) and the newly emerging "white ethnic movement" (Greeley, 1974; Stein and Hill, 1977; Weed, 1973).

Nathan Glazer (1975:202) reaffirms the theme of this essay in writing that social policies (involving housing, employment, busing) assume that ethnic groups are easily bounded and defined" and so uniform in the condition of those included in them, that a policy designed for the group can be applied equitably and may be assumed to provide benefits for those eligible. Thus he underscores the problem for makers of social policy where group boundaries and group homogeneity are uncertain.

At issue is the construction of social policies. Their goal should be to approach as closely as possible the philosophical ideal, "all men are created equal." A policy maker whose task is to promote social equality must understand social structures of ethnic groups in addition to the social structure of the society as a whole. While knowledge of the second convinces the policy maker that action must be taken to rectify social inequality, only an in-depth knowledge of the first will give him the means to achieve it.

Notes

1. For additional information on the phenomenon of ethnicity, see Depres (1975), Epstein (1978), Geertz (1963), Holloman (1978), Kuper and Smith (1969), Rubin (1960), and Shibutani and Kwan (1965).

2. For additional information on American ethnic groups, see Carlson and Colburn (1972), Greeley (1971), Howard (1972), Johnson (1976), Litt (1970), Makielski (1973), Mindel and Habenstein (1976), and Novak (1972).

3. For additional and comparative information on Greek-Americans, see Gizelis (1974), Kourvetaris (1971, 1976), Leber (1972), Papanikolas (1970), Patterson (1970), Politis (1945), Saloutos (1964), Schumach (1978), Stephanides (1971), Stycos (1948), Tavuchis (1972), Theodoratus (1971), Vlachos (1968), and Warner and Srole (1945).

4. See Gizelis (1974) for more information on Philadelphia Greek-Americans.

5. See Du Boulay (1974), Friedl (1962), and Sanders (1962) on the topic of contrastive male and female roles in rural Greece.

6. See Xenides (1922) for supportive viewpoint.

7. See Kourvetaris (1971) for information on Chicago's Greek-Americans.

8. See Kourvetaris (1976) on Greek-Americans in Detroit.

9. See Sanders (1962) on the distinction between rural and urban Greeks in Greece.

10. See Safilios-Rothschild (1967) on class position and success stereotypes in Greece.

11. A change in the Federal Immigration Policy of 1924 severely limited immigration from southern Europe until post-World War II.

12. See Patterson (1970) for a discussion of the unsuccessful among the oldtimer Greeks in Denver.

13. See the following works for information on affirmative action programs: Bureau of National Affairs (1973), McCune and Matthews (1975), U. S. Commission on Civil Rights (1973), and Williams (1977).

8

Social Equality and Infectious Disease Carriers

BARUCH S. BLUMBERG

Illness can have a profound effect on an individual's behavior and his social status. The carrier state for an infectious disease refers to a biological condition in which a person is infected with and "carries" a potentially disease-causing agent but is not himself, or herself, sick. This can raise unusual problems related to social equality, and these will be discussed, using the carrier status for hepatitis B virus as an example. To present this example, it will be necessary to give background information on "Australia antigen," hepatitis B virus (HBV), the diseases it is related to, and the carrier state for HBV (for a more detailed review, see Blumberg, 1977).

The Australia antigen was initially discovered during the course of a systematic investigation of inherited antigenic variation in human blood. This antigen was present in some individuals but not others and often led to the development of antibodies in patients transfused with the factor. In 1967 its association with acute and chronic hepatitis was shown. It subsequently was demonstrated that Australia antigen is on the surface of hepatitis B virus, the causative agent for many cases of acute hepatitis. The term hepatitis B surface antigen

This work was supported by USPHS grants CA-06551, RR-05539, and CA-06927 from the National Institutes of Health and by an appropriation from the Commonwealth of Pennsylvania.

(HBsAg) was subsequently used instead of Australia antigen. It is also directly related to several forms of chronic liver disease, probably including primary cancer of the liver, that is, cancer which starts in the liver (Blumberg and London, 1980). It was immediately realized that the test for Australia antigen, that is, HBV, could be used to identify asymptomatic carriers of the infectious agent. It was soon found that the "carriers" identified by this test could transmit hepatitis by blood transfusion. This led to the nearly universal practice in many countries of testing donors of blood used for transfusion in medicine and surgery, and this has resulted in a decreased frequency of post-transfusion hepatitis due to hepatitis B in many places (Senior et al., 1974).

It has always been clear that hepatitis B may be transmitted by means other than transfusion. Hence, consideration had to be given to the transmission of hepatitis from the carriers to others with whom they came in contact, i.e., family, co-workers, friends, etc. This has resulted in what may be an unprecedented situation in medicine and in society (Blumberg, 1976). The "carrier" is not apparently ill (although there may be in some cases detrimental, long-term effects of the virus on the liver), and he or she is not aware of differences from others until this occult HBsAg carrier trait is determined by a blood-screening test. The carrier is thought to be able to contaminate others with whom he comes in contact, particularly those with whom he is most intimate, although the risks imposed are not enormous and vary greatly from carrier to carrier. There are other examples in medicine in which person-to-person transmission occurs, but none of these is quite like the interactions of HBV and humans. For example, patients with leprosy, active tuberculosis, and other infectious diseases are ill and require treatment and are usually aware of their disease while, as already mentioned, the HBV carriers are not apparently sick. People can be occult carriers of typhoid, staphylococcus, and other agents, but a very small proportion of the population is tested to see if they carry these infectious agents, whereas, because of blood-donor screening, millions of people are tested each year to see if they are hepatitis carriers. Asymptomatic carriers of syphilis, gonorrhea, and other venereally transmitted organisms are in a somewhat similar situation to HBV carriers, but they transmit the disease primarily through sexual contact, and they can be treated, whereas hepatitis may be transmitted by a variety of mechanisms, and the carrier, once identified, cannot at present be treated.

At the outset it should be said that we do not have detailed or quantitative information on which carriers can transmit hepatitis or on the extent of the hazard they present. Carriers probably differ in infectiousness, but the characteristics which determine this are not well known. For example, it has been suggested that carriers who also have evidence of active hepatitis are more likely to transmit hepatitis than are carriers who have no evidence of disease. Theoretically, it would be expected that carriers with a relatively high concentration of the whole virus particles (which are thought to be the infectious virus particles of hepatitis B) would be more likely to transmit hepatitis than those without the whole virus Dane particles. It has also been stated that carriers who have the "e" antigen, an antigen in the core of the virus which can also be detected in the blood, are more likely to be infectious, and this hypothesis is being tested. It is clear that the appropriate epidemiologic studies have not been done, and much of the evidence for transmission of disease from asymptomatic carriers to others rests on anecdotal information. There are several published reports and some unpublished stories of dentists, physicians, nurses, and other health personnel who have transmitted hepatitis to patients with whom they came in contact. In at least one study of health personnel carriers it was concluded that they did not transmit hepatitis B (Alter et al., 1975).

Hence, we can tentatively conclude that under appropriate circumstances, some carriers can transmit disease, and at present it is not possible to identify those most likely to do so; although methods to do so may soon be developed. However, the risk is not enormous and is conceivably amenable to control measures. Independent of the actual data that are available, carriers are in many cases regarded as hazardous sources of disease. The myth is at present stronger than the facts and will probably remain so until more knowledge is available. There are few systematic studies of the social and personal consequence of being identified as a carrier. My colleague, Dr. Jana Hesser, has conducted a study to obtain information on the unusual and sometimes unfortunate consequences of identification as a carrier, and this is being prepared for publication.

We will now look at some examples (taken, in part, from Hesser's study) to see how detecting the carrier status has affected the lives of apparently well, asymptomatic individuals whose activities have been restricted presumably to benefit the health of others. Some carriers have been forced to leave their work as nurses, physicians, hospital and laboratory workers when they were identified as carriers. Recently, when visiting a country with a high frequency of carriers, I was asked if regulations should be passed to prevent the admission of carriers to medical school and to military officer training school. (I advised against this.) It has been recommended that nurses should be screened for HBsAg at the time of their entry into the profession and that those who are found to be carriers should be advised to "discontinue nursing as a career." This would be a difficult decision for an eager entrant into a profession for which he or she has prepared for many years. It has been suggested that health personnel should be routinely screened. Should this be compulsory or voluntary? Should a person be forced to take a test which might cause them to lose not only their present position but also the chances for other jobs in their profession? The problem becomes more dramatic when one realizes that carriers have always existed but that they were not identifiable until the development of the test; this special group of individuals was not known until the technical ability to identify them became available and commonly used. It is important to emphasize again that the risk of a carrier transmitting hepatitis to another individual by person-to-person contact must be relatively small. It is estimated that there are about one million hepatitis B carriers in this country. If they could transmit hepatitis very readily, then one would expect to find a much higher frequency of disease transmission than is known to exist.

Some additional problems have arisen that do not have obvious solutions at present. The frequency of hepatitis B is high in Vietnam, Thailand, and elsewhere in Southeast Asia, where large numbers of U. S. servicemen were stationed during the 1960s and 1970s. Hepatitis was one of the most common causes of morbidity in these military populations, and there is data to show that there is an increased number of carriers in the United States as a result of the experience in Vietnam (Blumberg et al., 1974). If it is a detriment to be classified as a carrier, is the military responsible for compensation of the carrier even though he or she is not obviously sick? The psychological impact of being classified as a carrier, possibly associated with loss of occupation, ostracism by fellow workers, and change in social habits, may be sufficiently detrimental to constitute a disability, and this will depend on the social attitudes which develop in the community. Carriers have been referred to as "lepers" in the medical press (Mosley, 1975) (the writer of this article condemned this practice); newspaper articles about carriers appear occasionally in the public press. A public attitude towards carriers is developing but, again, little is known about what it is. Hepatitis is not generally thought of as a calamitous disease, such as cancer, nor is it as benign as the common cold. How much do people fear it, if at all? How do they view social, professional, or sexual contacts with carriers or people with the disease?

Another potentially serious problem has developed in connection with adoption of children. HBsAg carriers are more common among Vietnamese children (about 6 percent or more) than among native Americans and north Europeans (about 0.1-0.5 percent). In a European country, Vietnamese orphans who were being considered for adoption were, among other things, tested for HBsAg. Some were found to be carriers. Should this be a deterrent to their adoption? Should this single test determine the course of a young life without any adjudication and without any real knowledge of the potential danger he or she might impose? Related questions have developed in relation to immigration of people from southeast Asia to North America and Europe.

Hepatitis and Artificial Kidney Units

The story of hepatitis transmission in renal dialysis units illustrates how the solution of medical and biological problems may change an ethical question and markedly reduce its impact.

After the development of methods for the detection of hepatitis B virus in the 1960s, it was learned that 50 percent or more of individuals treated in some renal dialysis units chronically harbored hepatitis B virus and had HBsAg in their peripheral blood. This was frequently accompanied by mild chronic hepatitis. Epidemics and sporadic cases of acute hepatitis, usually with jaundice and sometimes severe, have affected the medical, nursing, and technical staffs of these units, and deaths have occurred. Strenuous efforts were made to keep dialysis units free of hepatitis. In some places criteria for rejection of a patient for dialysis treatment included the presence of HBsAg in the patient's blood. If no other renal dialysis units were available in the community, then this was tantamount to allowing the patient to die. This, of course, raised an ethical question related to the selection of patients for treatment.

Studies of renal dialysis patients and staff followed, and the results from a unit in the Philadelphia region can be summarized (London et al., 1977). The staff were at a relatively high risk of developing hepatitis (10 percent per year of exposure) but a very small number became carriers (1 of 184). In all cases the staff who acquired hepatitis recovered. The dialysis patients themselves had a high risk of becoming chronic carriers. The patients who became chronic carriers did not usually develop acute hepatitis, although they did have mild abnormalities in their liver enzymes, and in some cases these abnormal chemistries persisted for long periods of time. In one study it was shown that chronic infection does not appear to increase mortality among the dialysis patients, nor did it adversely affect ability to receive transplanted kidneys. (It may, in fact, enhance it.) Hence, it appeared that there was a great risk to patients of becoming carriers, although this was not necessarily accompanied by disease. Staff, on the other hand, were at a much lower risk of infection, but they were more likely to develop acute disease.

Epidemiological control of hepatitis was established using the blood tests to identify carriers and prevent the spread of hepatitis in the unit. The carriers were treated on renal dialysis machines of their own or assigned to special units. In this instance the research and clinical advances helped to resolve the earlier ethical problem of how to deal with carrier patients.

Generalizations from the Hepatitis Carrier Studies

There are several generalizations which could be made from these experiences with hepatitis carriers. The problems that are raised are a result of a conflict between public health interests and individual liberty. When the risk to the public is clear and the restrictions on personal activities are not great, there is little problem in arriving at appropriate regulations. For example, the transfusion of blood containing hepatitis B antigen is a disadvantage to the patient recipient, and it has been stopped. Removal of "positive" units does not greatly decrease the amount of donor blood. On the other hand, the denial of the right to donate blood is not a great infringement of personal activity and the individuals concerned, and society has agreed to accept this moderate restriction. The problems raised by person-to-person transmission are more difficult. The extent of the hazard to the public is not clear since it is not (now) possible to distinguish definitely carriers who transmit disease and those who do not. On the other hand, if all carriers are treated as infectious, the hazards imposed on the carrier may be enormous, i.e., loss of job and ability to continue in the same profession, restriction of social and family contacts, etc. Hence, there is a very wide discrepancy between the potential but unknown benefits to society of restricting carriers and the real hazards to the individual.

What is clear is that for a very large number of carriers, the risk of transmitting hepatitis by person-to-person contact must be very small. Until we know more about this problem and have developed methods for distinguishing carriers who are likely to transmit hepatitis from those who are not, we should not cast all carriers into the same stigmatized "class." It is also likely that techniques for preventing transmission can be developed. Cases can be considered on an individual basis in deciding on their disposition, and encouragement given to sponsoring research which will solve the problems of detection and prevention of transmission.

On a broader level, the ethical issue is raised as to how much biological knowledge about individuals ought to impinge on the day-to-day contact between people. Is it appropriate to

regulate the risks inherent in people living together and interacting with each other? An issue has been raised with respect
to hepatitis because the test can be easily done and because
millions of people are tested as part of the blood-donor program. As a consequence of these identifications, this particular
group of carriers has been identified. There are other forms of
carriers, some of them potentially more hazardous (i.e., carriers
of staphylococcus and typhoid) who are not routinely tested
and, therefore, not placed at a disadvantage.

Ethical Issues and Basic Scientific Knowledge

A recurrent feature of our experience with ethical questions
has been the need to learn more about the biological, medical,
social, and psychological aspects of the problem; that is, more
of the *details* of the situation. It is now commonplace that scientific questions cannot be dealt with without a consideration
of moral and ethical consequences; and the opposite is also true.
An understanding of the ethical question often requires a more
detailed understanding of the scientific features of the problem.
Ethics and science are inseparable and they support each other.
For example, when faced with the ethical question of admitting
hepatitis carriers to artificial kidney clinics, it was necessary to
have technical and scientific information not available when
the question first arose. Additional research gave quantitative
information on the risk of infection to patients and staff, the
consequences of infection and other knowledge necessary to
arrive at a reasoned judgment on how to deal with the problem.
Our ability to obtain more detailed answers is often at the root
of understanding, changing, and sometimes "solving" the ethical
question; and it is this feature which makes the applications of
science to these problems so hopeful.

Bronowski and Mazlish (1960) in their discussion of Leonardo
da Vinci provided an insightful comment. "He discovered,"
they said, "that Nature speaks to us in detail, and that only
through the detail can we find her grand design." The ability
to collect detailed information is a peculiar strength of science
and its mode of operation. This talent can be used in obtaining
some of the answers needed for ethical questions. The specific

ethical problems which develop as a consequence of a medical practice are often unknown until the practice is established. It then becomes clear that new information is needed to deal with the ethical problems, and this information often derives from an understanding of how the specific medical question fits into the rest of nature. That is, the better forearmed we are with a broad understanding of nature, the more likely we are to be able to deal with questions as they arise; and this is an argument in favor of "nondirected" research early in the investigation of practical medical problems.

9

Societal Responsibility and Genetic Disease: Some Political Considerations

ROBERT H. BLANK

To what extent does a democratic society have the duty to take steps necessary to reduce the occurrence of genetic disease? If it has such a responsibility, at what stage does societal good preempt individual rights to procreation, privacy, and self-actualization? What responsibility does society have toward future generations, and is human genetic intervention justifiable on these grounds? While these questions obviously have no simple or absolute solutions, they are questions that must be examined in detail when discussing any attempts to intervene in the genetic composition of the population.

Facing such questions is even more crucial in light of the major advances in genetic technology which have occurred in rapid succession over the last decade, especially in the area of genetic screening and prenatal diagnosis. In utero diagnosis of affected fetuses through the use of amniocentesis and ultrasound techniques has been coupled with tests to determine carrier status of heterozygotes and identification of inborn errors of metabolism in newborns. These technical advances have been accompanied by changes in the attitudes and values of the public relating to abortion as an alternative to carrying a defective fetus to term. One result of the technological possibility of widespread screening for an increasing variety of genetic diseases has been to raise a number of new political and ethical issues to be faced.

Although many of the techniques for human genetic intervention such as gene therapy and surgery remain but theoretical possibilities, areas of immediate concern include: genetic counseling and screening, prenatal diagnosis, eugenic techniques such as sterilization and artificial insemination, and research on the mutagenic effects of various environmental factors. While this essay focuses on genetic screening and prenatal diagnosis, the other concerns above are intricately related and will be included in the discussion where appropriate.

While the goal of human genetic research is to reduce genetic disease, as prenatal diagnosis and screening techniques become more precise and inclusive, certain individual rights will be severely challenged. Not surprisingly, there is much disagreement over both the necessity and right of society to intervene in childbearing decisions. A major question is whether genetic screening should be considered within the realm of public health policy or remain a private responsibility? The assumption here is that eventually some trade-off between individual and societal benefits must be reached. A prerequisite of any decision in that direction is open and frank debate over the merits and disadvantages of each attempt at intervention.

The Social and Political Context of Genetic Intervention Decisions

While the presence of genetic disease certainly is not a new phenomenon, it is only within recent decades that we have come to have a basic understanding of the nature and extent of genetically related health problems. While knowledge of many multifactoral diseases such as multiple sclerosis, diabetes, and rheumatoid arthritis remains severely limited and the causes of many diseases continue to be unknown, understanding of many single gene and chromosomal disorders and an increasing number of metabolic disorders is advancing rapidly. Along with this understanding have come techniques designed to identify and, in several cases, treat genetically related disorders. Increasingly accurate and inclusive prenatal testing has been matched by more precise neonatal and adult screening techniques.

Along with these technological advances, value changes have taken place in American society which tend to facilitate the

acceptance and application of these new techniques. There is
evidence of support at least for the use of many techniques
that until very recently might have been rejected by large majorities. While still highly controversial, abortion of diagnosed
defective fetuses appears to be viewed as justified by a substantial
majority. In many sectors a more realistic view of genetic disease is held than ever in the past. While guilt might still accompany families with genetically affected offspring, society as a
whole appears to be more open and understanding. These
perceived attitude shifts have reinforced the acceptance of
technologies such as amniocentesis and certain screening procedures and have offered a more conducive atmosphere for
genetic counseling services. Although much opposition exists
in some groups, the government and a majority of the citizens
appear to be open to innovations in these areas.

During much of history, genetic disease has been a fact of life
with little human intervention possible in its reduction. While
negative eugenic attempts such as compulsory sterilization, laws
against certain types of consanguineous marriages, and premarital
blood tests for syphilis were enacted to reduce genetic disease,
their overall impact on the population has been questionable
(Reilly, 1977). Presently, however, in light of new diagnostic
and screening techniques, genetic disorders or even carrier status
can be identified directly with a high degree of accuracy. This
implies that what in the past was considered unavoidable, i.e.,
children with various genetic disabilities, now is avoidable in
many instances. The presence of these techniques and the promise of future advances in genetic technology necessitate analysis of the extent to which particular genetic disorders must be
considered within a public health context.

The question as to whether reduction of genetic disease is
a matter of public health is crucial as one considers compulsory screening or prenatal diagnosis. It seems logical that more
drastic mandated actions are warranted in situations where the
threat to public health is probable and immediate than where it is
not. While compulsory immunization for highly contagious
diseases has been accepted as a matter of public health, mandated screening for carrier status of various genetic diseases
appears to be a less immediate and viable concern to many

observers. Obviously, genetic disease is a threat to those directly or indirectly affected. However, genetic disease is not a threat to society in general as is a highly contagious disease. There is no danger that genetic disease will become epidemic. Although some of the arguments that support the deterioration of the gene pool appear plausible (Muller, 1959), there is little direct threat to immediate population or generations of the near future (Lappé, 1972).

It is imperative in any situation to scrutinize the nature and scope of those affected by any disease before placing it within a public health context. Responsibility of society toward these individuals or groups must be gauged, and their freedom of expression and choice must be balanced against some broader responsibility for society. Considerations in any attempt at human genetic intervention must be given to at least four elements. The four dimensions of responsibility include: (1) the affected persons, (2) the parents and family, (3) society, and (4) future generations.

Responsibility to the Affected Persons

First responsibility, it seems, should go to those actually affected by the disease. Although this seems to be an obvious assumption, there is a tendency by many to ignore this responsibility and focus instead on the rights of the parents or the needs of society. This is reinforced since in many cases the affected person is a fetus undergoing prenatal diagnosis or a hypothetical person of a particular union.

The responsibility to affected persons can relate either to those presently living with a genetic disorder or potential persons who would be affected if they are born. It seems that the responsibility toward those already living is to protect their well-being and to provide all possible means of minimizing their problems. The responsibility toward the yet unborn persons or more hypothetical potential persons that might result, for instance, from a union of two carriers, is much less clear and fraught with dilemma. Certainly, there is a societal responsibility to reduce the probability of any one individual being genetically defective. Does this mean, however, that any fetus so diagnosed should be destroyed? While from a societal standpoint such action eliminates the problem in that case, for some this is equivalent to murder. Generally, American society would accept abortion under such circumstances but, interestingly, not out of responsibility to the affected fetus but on some parental or societal basis.

Responsibility to affected persons becomes even more ambivalent when discussing screening for carrier status. Who then is the affected person? While carriers of the sickle-cell trait, for instance, might suffer some trait-related problems, certainly they do not have a disease. It is their potential or actual offspring which have a one-in-four chance of being affected. On this basis, does society have the duty to mandate the screening of carriers in order to protect potential affected persons? If it has the responsibility to make screening of carriers compulsory, does it then have the further duty to outlaw sexual unions between carriers based on the chance that some of their offspring might be affected to protect these potentially affected persons?

Implicit in such an argument would be the sacrifice that might result if the two carriers were not allowed to reproduce. In other words, once one focuses on potential affected persons, a dilemma arises in that probability dictates that any action taken to eliminate the union of two carriers will affect a larger number of potentially healthy than unhealthy children. On this basis alone, without taking into consideration the rights of the carriers, compulsory screening of carriers and its logical extension of prohibiting procreation in such instances would be illogical on public health grounds. Obviously, this does not preclude the screening of carriers on voluntary grounds or even the justification of mandatory screening based on some broader responsibility.

Parental Rights

In addition to consideration of the rights of affected persons and unborn or potential individuals, society must provide parents or prospective parents with adequate information upon which to make informed decisions. At some point, consensus must define if and when responsibility to the affected persons takes precedence over the rights of the parents to have or not to have children. This question without doubt is a difficult and sensitive one, and any determination of the boundary between parental rights and responsibility and concern for the affected person is bound to be controversial. Consequently, it is doubtful that any solution will be forthcoming. The extent to which society is bound to overrule the right to procreate in order to protect those affected or potentially affected by genetic disease, however, will be an evermore crucial question as more is understood about genetic disease. Obviously, the rights of the parents and those of their affected offspring will not always be at odds. Given the pluralistic value system in the United States, however, conflict is bound to be common, nevertheless.

In addition to the duty of society to minimize its interference in the decisions of individuals to reproduce as they desire are other concerns relating to responsibility towards individual citizens. Opposition to abortion on moral or religious grounds is intense and sincere and cannot be ignored as a central aspect in establishing any genetic intervention programs. Since most prenatal diagnosis techniques and goals center on therapeutic abortion as an alternative to carrying a diagnosed defective fetus to term, amniocentesis has become a target of those groups concerned with right to life. Any societal decision must take account of these moral dilemmas raised by intervention in the life process.

Additional areas of concern relate to individual prerogatives to privacy in the procreative process. Criticisms of amniocentesis as an invasion of privacy are strongly held by many, despite a general societal acceptance of the objectives of that technique. Questions relating to stigmatization of couples who choose to have children despite the high risk for their offspring and those who reject prenatal diagnosis or therapeutic abortion as alternatives must also be examined. Responsibility to parents as citizens with certain childbearing freedoms must be emphasized, although it cannot be the only consideration in any policy decision.

Societal Good

A third consideration in decisions concerning the application of various methods of human genetic intervention relates to the needs of society at large. In addition to the indirect influence of societal values on all policy decisions is the more direct impact of the costs society is willing to bear to care for those with genetic disease or to implement genetic intervention programs. This question is certain to become more crucial as resources are constrained and competition for these scarce resources becomes more intense. Recent moves to restrict taxing powers of the states are but one evidence of the movement toward less government spending. Programs with low visibility and narrow constituencies are most likely to suffer. While care of the mentally retarded at present is less than adequate, it seems unlikely that the situation will improve.

Also, as the population increases and the proportion of the working members decreases, pressure to restrict support for the "genetically defective" might increase. One result might be increased efforts to reduce or eliminate the incidence of genetic disease by introducing mandatory screening coupled with restrictions on procreation by carriers. Compulsory diagnosis of high-risk fetuses and implicit or explicit pressure to abort fetuses diagnosed as having genetic defects might also follow under such circumstances. While there certainly will be pressures in the opposite direction to protect the rights of parents and concerned others, the rapid and pervasive proliferation and acceptance of cost-benefit analysis indicates a long and intense battle. Thus, while it appears unlikely that compulsory genetic intervention programs could be justified from an individualistic standpoint, societal pressures on the distribution of limited resources might preclude certain options in the future, especially as technology offers inexpensive, effective, and less intrusive means of testing. While a public health justification might be projected under such circumstances, at base would be a monetary decision.

The most common orientation for political decision making today is the public interest. Generally, the public has come to refer to either the majority or some subset of it. The public interest on any issue is at most the interest of some majority of those involved in that specific issue. It is seldom if ever some objective good for society at large. As various "publics" conflict, it becomes difficult to demonstrate the existence of benefits which would accrue to all members of society.[1] The concept of public interest then is much more restrictive than the traditional concept of the common good. As a moral justification for public policy, especially on the issues described in this essay, the public interest appears to be quite limited.

The liberal concept of special or attentive publics further confuses decisions in the area of genetic research. Much of the current pluralist theory accepts the notion of intense minorities and their rights to protect their interests despite the implications for society as a whole. In terms of genetics, interest group pluralism might limit the public to a small proportion most directly affected by the various techniques under consideration, i.e., those with genetic diseases. Does society have the right to limit certain research which might benefit only a minute fraction of the population, and does that minority have the right to lobby for such research? In many states, primary impetus for genetic screening programs for PKU, Tay-Sachs, and sickle-cell anemia came from organized interests of those directly affected. They functioned not as common good advocates but as a special interest group with specific demands for aid in preventing small numbers of future citizens from sharing their experience.

This brief discussion demonstrates the difficulty in defining societal good itself. While responsibility to society certainly is a major consideration in policy decision relating to genetic intervention, it, too, fails to provide objective standards for making such decisions. The problems in defining the public have always been overwhelming. In an area where subjects of the research many times are fetuses, embryos, and potentially single genes, this problem is even more imposing.

Responsibility to Future Generations

It is most difficult to include generations beyond the near future in policy decisions within a public interest framework, since most decisions tend to be made in response to immediate concerns. However, current emphasis on environmental pollution and depletion of natural resources has created an awareness of our responsibility to future generations to pass on a world where there is at least a reasonable chance of confronting successfully the problems left by twentieth-century humans. Some have argued that this concern should be directed, not only to the obvious cases of population size and the environment, but also to medical and genetic technology. Kieffer (1974:86) argues that at a minimum, the living have an obligation to refrain from actions that endanger future generations' enjoyment of the same rights now assumed. Proper moral concern is not limited to the near neighbor but also the distant neighbor in space and time. We must be aware of any processes which might be irreversibly harmful for future human life.

In terms of genetic intervention techniques, responsibility to future generations must become an integral aspect of any policy decision. The decisions made in the next decade are likely not only to modify our conceptions of humanhood, but also potentially to alter future individuals and the prospects of continued survival of the human race. We cannot afford either blindly to ignore the opportunities presented by genetic technology, or actively pursue human genetic intervention without including futuristic considerations. Many of the recipients of the benefits of today's research are tomorrow's citizens. Unfortunately, any harmful effects of such research might irreversibly affect those same generations. Policy decisions made now, therefore, must consider, to the maximum extent possible, such concerns.

A central question is the extent to which future generations must be considered. How much effort should be made to include the interest of those yet unborn? Who should speak for the future? Do we have a moral obligation to deny ourselves certain advantages now in order that those who are not yet born may live better (Kieffer, 1974:180)? In the areas of environment and population, questions are at least being asked. In the area of genetic technology, the question is much more complex since its goals are less clear and the implications more complicated. For instance, certain types of genetic engineering might indeed reduce the suffering of parents and children who would otherwise have been born with birth defects. Even if this potential is successfully met, however, there is the possibility that mutations formed by such a process might have adverse effects on generations of the distant future. Do we ban research on this basis or simply minimize the possibility of adverse long-range consequences?

Another position is that first obligation is to more immediate generations and not to some potential populations. Just as the present generation has had to adjust to the effects of decisions made by past generations, i.e., slavery, exploitation of natural resources, etc., so future generations must adapt to decisions made now. Although most would agree that no decisions consciously should be made that would endanger the rights and survival of future humans, it could be argued that primary responsibility lies in those now living and their immediate offspring. This position encourages only minimal restraints upon actions based solely on the fate of those in the distant future. According to Golding (1968:457), for instance:

> It is highly doubtful that we have an obligation to establish social programs that would secure a "good life" (prevent the undesirable, promote the desirable) for the community of the "remote" future. The conditions of life then are likely to be so different from any that we can now imagine that we do not know what to desire for them.

This second position seems to be less tenable under conditions faced in the 1980s simply because today's actions more than ever before might constrain greatly the alternatives open in the future. Decisions made now might inalterably limit or expand the decisions of all who follow. It is this possibility that should make one more aware of the potential of current technology. Past technologies polluted the waters, and we have to live with that fact. We can either reverse that trend or accelerate it. Future generations may not have that choice if we make the wrong decisions. Through intervention in the genetic composition of the human species, the possible harmful consequences to those of the future must be considered along with the possible benefits of such intervention. Certainly we cannot make decisions for those in the future, but we must be aware that each has broad implications on future alternatives.[2] Unfortunately, it is not possible to determine whether implementation of a specific genetic intervention program is desirable from a future-oriented standpoint since the long-term consequences cannot be explicated from current knowledge in many cases.

Societal Considerations vs. Individual Rights

Gustafson (1973:102) argues that the "major persisting matter of moral choice is whether preference should be given to the individual or to a community." There are many ways in which this tension between individuals and society can be stated. One can talk about rights of the individual vs. costs or benefits to the community; or the rights of the human race to survive vs. the rights of a mother to bear a defective child. Whatever the distinction, most resolutions of genetic dilemmas center on this conflict between the individual and the good of the community. Rights of individuals must be weighed against the rights of society. Callahan (1973:240) sees little chance of a happy balance and contends that all solutions are bound to be only temporary.

Fletcher (1974) sees the conflict reduced basically to a sanctity-of-life vs. a quality-of-life ethics, and he opts for the latter. His "situational ethics" places emphasis directly on the principle of proportionate good. He contends that one must attempt to compute the gains and losses which would follow from several possible courses of action (or nonaction) and then choose the one which offers the most good. The common welfare has to be safeguarded by compulsory control if necessary, according to Fletcher (1974:180): "Ideally it is better to do the moral thing freely, but sometimes it is more compassionate to force it to be done than to sacrifice the well-being of the many to the egocentric rights of the few." Fletcher favors compulsory controls on reproduction if they are needed to promote the greatest good for the largest number.

Taking an opposing position, Beecher (1968:133) argues that society exists to serve man and not vice versa. Therefore, individuals have certain inalienable rights which cannot be taken from them by the state. Paul Ramsey (1975:238) contends that "parenthood is certainly one of those 'courses of action' natural to man, which cannot without violation be dissembled and put together again. . . . " He also defends vigorously the rights of individuals to unique genotypes and sees germinal engineering as immoral (1975:235).

In examining the societal and individual costs of genetic disease, Lappé (1972:425) minimizes the social costs. He sees the burden resting mainly on the family, not society. He also dismisses the genetic load or genetic burden concept as well as Muller's (1959) theory of social cost. Lappé (1972:420) is disturbed by the current advocacy of societal intervention in childbearing decisions, denial of medical care to the congenitally damaged, and sterilization of carriers. He contends that society has no right to intervene in childbearing decisions except in very rare exceptions. Procreation should be the choice of the parents, since it is they who bear the burden of deleterious genes, not society.

The case of society's concern for the genetic welfare of the population and its rights in imposing sanctions on the individual rests on a yet unproven "clear and present danger of genetic deterioration" according to Lappé (1972:425). Callahan (1973) agrees that an affluent society should be able to absorb the costs. Parents should have the right to bear defective children even if the social costs of this freedom are high. He contends that it is difficult enough to resolve conflicts among individual rights, much less between individual rights and some nebulous greater social good. On the other hand, parents should not be forced to bear defective children.

While these two sets of positions provide a framework within which to analyze human genetic intervention policy, they result in a standoff due to their conflicting assumptions and their exclusiveness. Those emphasizing societal rights fail to account for the rights of the various individuals discussed earlier and for future extensions of the present population. Those who focus on individual rights of the parents overlook, or at least minimize, corresponding rights of potentially or actually affected persons and of others in the community. When examining specific applications of screening and prenatal diagnosis in the next section, one must keep in mind the complex interrelationships between the affected individuals, the parents, society, as well as future generations. Initiation of any genetic intervention program rests on full consideration of all four dimensions of responsibility.

Genetic Screening and Prenatal Diagnosis Applications

Although all genetic screening applications are similar in their objective of reducing genetic disease, the vast array of genetic disorders requires many different approaches and techniques. Some screening is aimed at the general population, while others are targeted at selective high-risk populations. Screening can also be conducted at various life stages. While prenatal diagnosis of the fetus is central to screening for Down's syndrome and other chromosomal abnormalities, PKU screening is conducted in newborns. Other metabolic disorders are most effectively screened in adolescence, while detection of carriers of recessive genes is best accomplished in young adults. This section of the essay briefly discusses some major screening approaches and summarizes current policy applications.

Prenatal Diagnosis: The Techniques

The most controversial form of screening entails the prenatal detection of genetic disorders through the use of amniocentesis followed by abortion of the defective fetuses.[3] Although amniocentesis and other screening techniques are presently limited, Peter (1975:202) suggests that eventually screening for all metabolic and chromosomal disorders might be conducted from amniotic fluid cells, therefore making routine the monitoring of all pregnancies.[4] Amniocentesis is the procedure by which fluid is withdrawn from the amniotic sac and then analyzed through a twelve-to-eighteen-day culturing process. Chromosomal as well as biochemical tests can be run on the amniotic cells. A relatively recent innovation (1966), it is already quite effective, with an error rate of between six and seven cases per 1,000. Despite some concern over the risk of conducting amniocentesis, it is quite safe; according to Milunsky and Atkins (1975:222): "In our experience with amniocentesis performed by many obstetricians with varying experience, the true risk of the procedure for causing fetal loss is probably less than 0.5 percent. No maternal mortality has occurred and maternal morbidity encountered has been minor."

Very recently ultrasound techniques have been performed prior to amniocentesis in about 10 percent of the cases, especially when twin pregnancies are diagnosed, in order to locate the placental position and diminish the risk of blood taps. As ultrasound becomes more commonly utilized, the risk of amniocentesis will be reduced further. Research now being conducted to isolate fetal blood cells in the mother's bloodstream might some day minimize the need for the physically painless but sometimes psychologically difficult amniocentesis procedure and necessitate only a blood test of the mother.[5]

At the center of such prenatal diagnosis is the role of the state. Milunsky (1976:58) argues that, while a voluntary program of amniocentesis followed by voluntary abortion of genetically defective fetuses is most equitable, evidence infers that such an approach is not effective. While a mandatory program of amniocentesis (most probably based on maternal age and other risk factors) followed by voluntary abortion would increase the effectiveness of the preventive program, serious ethical and legal questions make this approach difficult to accept. Legal questions of invasion of privacy and religious freedom arise, and at least one observer (Friedman, 1974:92) contends that mandatory amniocentesis for any reason would be challenged on grounds of violating the Fourth Amendment proscription of "unreasonable search and seizure." Any efforts at mandating prenatal diagnosis are likely to fail since, presently at least, there is no evidence that the state has any compelling interest to warrant intrusion into the procreation process on genetic grounds. While prenatal diagnosis should be available to those who desire it, any effort to mandate such procedures has little support.

The very presence of the ability to diagnose genetic diseases prenatally results in an ethical dilemma, centering on the attitudes of society toward those who bear children with condition which could have been diagnosed if the parents had desired.

Callahan (1973:248ff) sees a potential danger to the right of self-determination, for instance, if such screening becomes routine and therapeutic abortions automatic. Even if the decision of the parents is voluntary, this precedent would establish a "causal logic" in bearing a defective child. By giving the parents "freedom of choice" through advances in screening, we are also making them responsible for the choices they make. Callahan (1973) argues convincingly that it is only a short step to distinguishing between responsible and irresponsible choices under such conditions. The social pressure on parents to make the "responsible" choice might be powerful. Although they would, in principle, retain the right to carry a diagnosed defective fetus to term, it is not difficult to imagine the inherent pressures on them to abort for the "good of society" and the potential child. By providing full, informed freedom of choice, then, advances in prenatal diagnosis techniques might result in the abrogation of such freedom.

Prenatal Diagnosis of Chromosomal Disorders

Recent advances in chromosomal analysis, such as production of fluorescent chromosomes and computerized karyotyping, promise more efficient diagnosis of chromosomal disorders. Identification of an increasing number of biochemical defects through a single test will extend the use of amniocentesis to diagnose an even wider spectrum of autosomal dominant and recessive disorders. As the per-unit cost of approximately $250 to $600 for amniocentesis decreases, there is expected to be increased pressure on doctors and patients to utilize prenatal diagnosis to monitor high-risk pregnancies. One target group is certain to be women who are thirty-five and over, since the risk for virtually all chromosomal disorders rises rapidly with maternal age.

The most frequent and well-known chromosome abnormality is Down's syndrome. Although this syndrome can be inherited through translocation, the most common form is trisomy 21, where there is an extra chromosome for the twenty-first pair. Trisomy 13 and trisomy 18, although much rarer and usually fatal in very early childhood, follow a pattern relative to maternal age similar to Down's syndrome. The estimates below indicate the importance of maternal age in chromosomal abnormalities.

Mother's Age	Down's Syndrome	Any Chromosomal Abnormality
below 30	1/1500	1/750
30-34	1/1000	1/500
35-39	1/300	1/150
40-44	1/100	1/50
over 45	1/40	1/20

The most significant aspect of these figures is that resources consciously can be directed at the high-risk group of women over thirty-five or forty, depending on the availability of funding and equipment. In Boston, for instance, prenatal genetic studies are routinely recommended for women over thirty-five (Milunsky and Atkins, 1975:231).

This same process can be used to diagnose sex-linked disorders. Such disorders are usually found in males since they are expressed in the absence of the second X chromosome. Klinefelder's syndrome (XXY), Turner's syndrome (XO), and many other sex-chromosome abnormalities can also be identified, though in most cases no treatment is possible and screening all fetuses would be necessary to detect these disorders. Certain X-linked familial diseases such as hemophilia cannot be identified in utero. All that can be done in such cases is to identify the sex of the fetus. If the fetus is a female, it will not have the disease,[6] but if it is a male, there is a fifty-fifty chance and abortion is offered as an alternative.

One form of chromosomal abnormality which has become the center of screening controversy is the presence of an XYY complement. Early statistical data demonstrated a high incidence of this complement among prisoners. When it was found that XYY "supermales" were overrepresented in prisons and that some tended to express antisocial behavior (in addition to being tall and of a lower IQ than average), efforts were made to establish a test program to identify young boys with this genotype. Although there is much disagreement over what XYY means and no causal link has been demonstrated between this genotype and aggressive behavior, much controversy has raged since the first identified XYY in 1961 (Dershowitz, 1976:63-71).

Although mass screening for XYY has never been carried out, the question as to what to do with such information which becomes available in any prenatal diagnosis persists. Should parents with a XYY son be told the truth under these circumstances? There is at least some probability that their knowledge of this might be self-fulfilling. They might be looking for the predicted behavior, with the resulting stigmatization of a healthy normal boy. Another question is whether or not an XYY complement should be grounds for dismissal of criminal proceedings. In other words, if XYY is really linked to antisocial behavior, can society treat that person as a criminal? Should people with abnormal chromosome complements be sent to prison at all if they are genetically predestined to behave in a particular manner? Prior to 1961 this was of no concern. Again, increased knowledge presents us with new dilemmas.

One additional dilemma arising out of chromosomal analysis is the question of sex predetermination. Since sex is easily identified through amniocentesis, it is obvious that technically one might abort a fetus simply because it is not the desired sex. Although most doctors and genetic counselors reject the use of such procedures and refuse to disclose the sex of the fetus unless an X-linked disease is involved, there is little to stop those less ethical from disclosing such information. Certainly, the use of prenatal diagnosis and abortion to identify and terminate those fetuses with the "undesired" sex pushes to the limit even the most liberal attitudes towards abortion. Still, the question remains as to whether parents have the right to such information and to utilize it as they desire.[7]

Neonatal Screening: Phenylketonuria

While there are at least ten inborn metabolic disorders tested for in newborns in various states, Phenylketonuria (PKU) screening serves as a prototype due to its scope and early legal status. PKU is an autosomal recessive disease caused by inactivity of the enzyme phenylalanine. This metabolic defect results in severe mental retardation if dietary therapy is not begun soon after birth. With early detection and proper diet low in phenylalanine content, mental retardation can be averted. Furthermor it has been demonstrated that the diet generally can be discontinued by about age six without adverse effects (Reilly, 1975: 326).

In the early 1960s Robert Guthrie developed a bacterial inhibition assay test that could be automated to provide for mass screening of PKU. Since that time, the test has been improved and multiple tests for other metabolic disorders have been used in combination with the Guthrie test (Guthrie, 1973: 229-30). Although early PKU testing met serious problems with false-positives, quality control has increased, and new procedures have been adopted to minimize, though not eliminate, this possibility.[8]

The history of PKU screening legislation has been a short but an active one. Within two years after passage of the first mandatory testing program in 1962, thirty-one states followed suit. By 1970, forty-two states plus the District of Columbia passed PKU legislation. About half of the states now provide testing for other less common metabolic errors. Although the specifics of PKU screening legislation vary, most states mandate the testing of newborns.[9] A majority of these states, however, provide for exemption on religious grounds.

Reilly (1975) is critical of the legislation written by many states on several grounds. First, most of the states do not provide guarantees of genetic counseling to parents of affected children, nor do they provide for counseling of the affected person at an appropriate age. Second, only a handful of the states fund programs of public education or provide followup testing. Also, few state laws consider the confidentiality of test results or specify the manner in which such data should be stored, if at all. Reilly (1975:334) concludes that "the passage of shoddily written genetic disease screening laws" suggests the need for model legislation drafted by a committee of concerned individuals. This would help ensure "comprehensive and reflective treatment of medicoethical issues by a panel of experts that no single group of state lawmakers could amass"

(Reilly, 1975:336). The passage of comprehensive federal legislation to combat genetic disease appears to be increasingly
likely.

Despite some technical testing problems and the difficulties
of applying many state laws, PKU screening of newborns is the
most widespread form of genetic screening. As such, it serves
as a model for other screening programs. It is unlikely, however, that the haste in passing PKU screening legislation will be
duplicated for other diseases. One reason for this is that legislators are less likely now than a decade ago to commit funds
to untested health-related programs. Another reason is that
among all the screening procedures, PKU is probably the least
controversial because it results in the treatment of the disease.
Most important, the stigmatization and sensitivity of PKU
screening is more limited than for those programs where the
target population is both well defined and already the subject
of discrimination.

Carrier Screening Programs

A much more controversial type of screening than testing
for inborn metabolic disorders is screening for carriers of autosomal recessive diseases. The goal of such a program is to identify carriers of particular genetic traits and educate them as to
their chances of having affected offspring, hopefully thereby
reducing the number of children born with genetic diseases.
This objective might be accomplished by discouraging them to
reproduce or by providing prenatal diagnosis, if available for
that particular disease. Obviously, this type of screening is more
intrusive into the procreation process and provides an excellent
example of the potential conflict among parental rights, societal
needs, and the rights of the potentially affected individual.
Also, since the carriers do not have the disease, the screening
process is indirect and not beneficial to the health of those
screened, but rather to potential persons who have a one-in-four
chance of having the disease if they are conceived and born.

Carrier identification produces another dilemma of screening. It has been demonstrated that for many genetic diseases
such as Tay-Sachs, sickle-cell anemia, and thalassemia, the

incidence among particular ethnic groups is significantly higher than the population as a whole. Economically and logistically, screening for such high-risk groups is most practical, while screening the entire population would not be feasible because of the low overall incidence of carriers. However, it is specifically these diseases that result in stigmatizing certain groups. When these groups already are the target of discrimination socially, identifying them as carriers of a genetic disease can be devastating. This is especially the case in screening for the sickle-cell trait, since there is little medical value of detecting the carrier state unless it is used in conjunction with prenatal diagnosis.

Sickle-cell Anemia. Sickle-cell anemia is an autosomal recessive disease found primarily in American blacks, of which approximately 10 percent are carriers of the trait. Approximately 50,000 persons suffer from the disease, which can lead to early and painful death, but also might allow for a relatively normal life span. The cost of care for sickle-cell anemia is high, and since there is no well-tested prenatal diagnostic test available at present, selective abortion is not always an alternative. The only option is to identify carriers and counsel them on the risks of having an affected child if both parents are carriers. Unlike PKU, no treatment is available. Screening for sickle-cell anemia, therefore, is an extremely limited health care approach at present.

The early 1970s saw a great amount of interest in sickle-cell legislation. Both at the national and state levels, politicians embraced the concept of attacking this long-neglected disease and in some cases rushed into passing screening programs that were haphazard and poorly planned (NAS, 1975:117). In many statutes, little distinction was made between the carrier status and the disease, and much unnecessary anxiety and confusion resulted. Education of the public in some cases came after, rather than prior to, screening and even this generally was insufficient. Despite the fact that the first laws were often written and sponsored by black legislators, the repercussions in the black community were not anticipated and the support of the community was not ensured. Opposition to mandatory screening was strong, but voluntary programs also came under attack, primarily because of the stigmatization attached to being a carrier of this highly selective disease. Fear of very real discrimination by insurance companies and employers accentuated this opposition at the same time other states were anticipating adopting sickle-cell legislation.

At present, about one-third of the states offer screening for sickle-cell anemia. Legislation in a handful of these states makes screening mandatory. The National Sickle Cell Anemia Control Act passed in May, 1972, allocated $85 million over a three-year period, to be used for screening and counseling. Since one requirement of the act is that the states administer screening programs on a voluntary basis in order to qualify for federal funds, pressure is on the states to comply. This, combined with opposition from internal sources, resulted in Massachusetts's decision to amend in 1974 its earlier compulsory legislation to provide sickle-cell testing on a voluntary basis. The present state of knowledge about the disease and its sensitive nature certainly favor voluntary screening based on adequately funded education programs. They also require full support of the target population if they are to be successful.

Tay-Sachs Disease. Just as the sickle-cell screening experience offers a good example of how not to conduct carrier screening, the efforts to screen for Tay-Sachs disease appear to be a model for similar future efforts. Tay-Sachs disease is rare in the general population, but like many of the over 400 autosomal recessive disorders already known, it occurs more frequently in a specific group, in this case Ashkenazi Jews. A most unfortunate aspect of Tay-Sachs is that the baby appears normal and healthy for the first six to eight months. Gradually, however, the nervous system degenerates and death results by age four or five after a long and painful confinement. Paralysis, blindness, and severe mental retardation are common, and the care approaches $50,000 per year. The emotional toll of watching one's apparently normal baby deteriorate, of course, is heavy.

Unlike sickle-cell anemia, both carrier detection and prenatal diagnosis techniques through amniocentesis are available for Tay-Sachs. Prospective prevention of Tay-Sachs disease is feasible since:

1. It occurs primarily in a specific population where selective screening can be effective.
2. There is a simple, accurate, and inexpensive carrier detection test.
3. The condition can be detected early in pregnancy, enabling selective abortion if the parents desire.

While Tay-Sachs offers a feasible condition for mass screening because of these factors, it also reflects the problems posed by all genetic programs. Unless the program is conducted with care and utmost concern for human values, it, too, can be counterproductive. The success of Tay-Sachs screening in several communities, such as Washington D.C., however, offers hope for similar screening of other diseases.[10]

Carrier Screening Problems. In screening for carriers, one is no longer dealing with a treatable disease such as PKU. In most cases, there is no disease at all but rather heterozygous carriers of the particular genes. The stigmatization attached to carriers causes much misplaced apprehension and bitterness, even under the best of circumstances. Therefore, screening of this type presents an even greater dilemma than mass PKU testing programs. Any state considering carrier detection programs must be certain to have the full support of the affected population and must be willing to expend significant resources on education and publicity before the actual screening begins. Even then, screening will be controversial because it restricts certain individual prerogatives, either implicitly or explicitly.

The choice of approaches of voluntary screening programs in a community must be made early. One can either attempt to screen a major proportion of the population at risk or make the service available to those who learn of the service and come in on their own. In either case, public education is vital. Due to limited facilities and resources for some types of screening, a decision must be made as to whom should be screened. Adequate counseling services and followup procedures also must be provided in any screening program. One of the shortcomings of many PKU screening programs has been the failure to provide extensive followup counseling. Sickle-cell anemia programs

have been plagued by a shortage of trained counselors as well
as adequate followup procedures for those testing positive.
Efforts must also be made to inform and educate the medical
community and to seek its cooperation in any screening efforts.
   While mass screening is most effective when a well-defined,
high-risk population is identified, the mere concentration of the
disease in that population can easily lead to stigmatization and
discrimination. The social costs of a poorly planned and insen-
sitive program can be most disastrous, due to the nature of the
subject. Attempts at sickle-cell anemia screening programs in
several states and localities have failed because the screening
was conducted without full comprehension of the social and
personal anxieties produced. One must understand that the
mere testing of a person can be threatening. Only education and
counseling can alleviate this threat and reduce anxiety.

Implementation of Genetic Intervention Programs: Compulsory or Voluntary Compliance?

   An issue central to all potential screening and eugenic programs is whether society should mandate a particular action for
reasons of public health or some other societal goal or whether
any action should remain voluntary. While this is commonly
presented in strict "compulsory versus voluntary" terms, it is misleading to discuss the implementation of such programs as a
dichotomy since even a voluntary program has an inherent
coercive aspect if accepted by society. The choice of policy
options is better represented by a series of approaches ranging
from purely informative to compulsory, depending on how
compliance is effected. The most coercive approach is to compel
screening by law. This could apply to the entire population or
certain high-risk groups as determined by lawmakers. The
severity of the compulsory aspect depends on the penalties
assessed to noncompliance. Obviously, if criminal penalties are
provided, compliance will increase, although public acceptance
of such laws will be difficult to achieve in all but extreme circumstances.
   An intermediate approach to genetic screening or eugenic
attempts is to construct incentive programs of greater or lesser

intensity. Positive tax incentive programs have long rewarded marriage, encouraged home ownership, and influenced family size. While negative financial incentives are less obvious, limits on the number of children eligible for AFDC payments and restrictions on the distribution of federal funds for prenatal diagnosis and abortion might have a direct influence on genetic screening programs. Another type of incentive program could be the provision of particular services without cost. Attempts to provide broad access to family-planning services and genetic-counseling clinics are prime examples of the incentive approach to implementation of specific principles. While such programs are less directly coercive in nature, they might be used to achiev the same objectives as legislation without the onus of the law. Their success largely depends on the extent to which they increase compliance to the predetermined goal.

A last approach to implementation of genetic intervention programs is simply to provide citizens, or those in high-risk groups, with specific information upon which to make an informed choice. While experience demonstrates that this approach is the most difficult with which to achieve broad compliance, the lack of direct coercion and the resulting impression of free choice makes it most appealing in those cases where there is no serious and direct threat to the health of the public. Despite its appearance as unobtrusive and the assumptio that it guarantees free and informed consent, this approach too includes a degree of coercion. The mere dissemination of available information carries with it at least implied pressure to conform with some standard defined by society. Therefore, while this final approach minimizes such pressures, it would be foolish to state that any approach is totally objective. The choice then becomes one of the degree and type of coercion utilized to assure compliance, as well as the means employed to achieve the policy goals.

Genetic Screening Programs

Due to the recency of genetic screening and diagnosis progran and the lack of standard legislation across the states, implementation of these programs tends to be fragmented and varied fron state to state and program to program. While most attempts currently are based on incentives and voluntary compliance, two types of laws have been established which mandate particular screening procedures. The first is aimed at detection of affected individuals so that treatment can be offered. The best example is PKU screening of newborns and similar programs for detecting metabolic disorders. The second type of mandatory screening program is designed to identify carriers of recessive deleterious genes and inform them of the risk of bearing children with genetic diseases if their spouse is also so identified. The most obvious example of this second type is compulsory screening for sickle-cell anemia.

Screening of the first type, when combined with treatment of affected individuals, might be justified as a public health measure and as a means of saving public funds. The compulsory screening element might be supported by the *parens patriae* doctrine that the state can act to protect those that cannot protect themselves (NAS, 1975:189). When specific conflicts with religious freedom must be resolved by the courts, mandatory screening of this type is much less controversial than compulsory screening for carriers.

There is little evidence that screening for carriers can meet either the public health or the paternalistic grounds of justification.

> We have been impressed with the effectiveness of the voluntary approach and we believe that the psychologic, political, and moral dangers of legislating human genetic testing far outweigh the potential medical benefits. . . . As a practical matter, legislation alone cannot solve genetic problems. (Kaback and O'Brien, 1973:262)

Since this type of screening contains implications for influencing reproduction decisions, it is effective only if the carriers refrain from having children or make use of prenatal diagnosis followed by selective abortion where appropriate. Each of these steps is controversial and, ethically as well as legally, questionable at best. There appear to be few benefits and many costs to mandatory screening of carriers. It is here argued that any screening programs for carriers of autosomal recessive diseases be predicted on the assumption of voluntary compliance Certainly, an adequate education program and the availability of services to all citizens is a solid base from which to proceed.

Although screening for PKU and related diseases appears to be less troublesome, a basic objection to all mandatory screening is its potential interference in "respect for individual choice in child-rearing matters" (NAS, 1975:189). While screening is intended to provide accurate information upon which informed choices can be based, there is some concern that it should be up to the individual as to what information is sought. The know-ledge of adverse information itself might be a "powerful impetus toward action," in effect offering very little choice. Callahan's (1973) concern for the casual logic of voluntary decisions is most appropriate in mandatory screening, since once the information is available, free choice is constrained. In other words, the only stage at which free choice can be assured is if the decision to screen is voluntary. If not, all succeeding choices are at least in part bound by the knowledge of the screening result.

Although it might be assumed that it is preferable to know that a fetus is affected with Down's syndrome early in pregnancy, it might, similarly, be argued that only the woman should be able to make that decision without coercion since, if diagnosed positively, the pressures for an abortion might be counter to her sense of personal morality. While such intrusion into private childbearing decisions is much less drastic, for instance, than sterilization, the constitutional questions of mandatory screening are yet to be answered. It is probable that screening where treatment is available, which does not directly infringe on the right to bear the child, is on a more solid foundation ethically than prenatal diagnosis and screening of carriers where alternatives are limited even with the screening information.

The extent to which mandatory screening programs can withstand legal and social challenges, then, will most likely depend on the procedures and objectives of each program. Screening conducted with minimal risk to those being screened, which results in treatment, is more easily defended than that for untreatable metabolic disorders. Screening for recessive gene carriers, while offering more informed decisions about marriage and reproduction, at this time should remain voluntary. At the same time, however, education of the public and screening facilities should be made available on a voluntary basis. Tay-Sachs programs should serve as a model for sickle-cell programs.

Whenever there is doubt, it is most crucial that genetic screening programs and prenatal diagnosis be voluntary.

> As a general principle, we strongly urge that no screening programs have policies that would in any way impose constraints on childbearing by individuals of any specific genetic constitution; or would stigmatize couples who, with full knowledge of the genetic risks, still desire their own children. (Lappé et al., 1972:1131)

Lappé et al. (1972), contend that while vaccination against contagious diseases and premarital blood tests are sometimes necessary to protect public health, there is currently no public health justification for compulsory carrier screening legislation to prevent genetic disease.[11]

The Need for Public Education

Despite an increased interest in genetic technology by the press and concerned segments of the public, genetic disease continues to be an area of ignorance for a major proportion of the public. Within the context of new screening legislation and the awareness of the rapid advances in genetic technology, most lawmakers tend to be uninformed and only marginally concerned. This is not unexpected because of the heavy workload of most legislators and the presence of more immediate matters. It is difficult under such circumstances to convince lawmakers to provide public funds for education concerning genetic problems and/or programs. Even in those states with energetic screening programs, money for educating the public constitutes only a minor fraction, if any, of the program cost. Instead, the funds go almost exclusively for the actual cost of the tests which, being more visible and easily measured, appear to be funds well spent. As the advances in genetic technology continue to accelerate and the technical possibilities for human genetic intervention multiply, it is imperative that public education be made an integral part of any attempts to apply the technology. This need centers on two aspects of education. First, there must be more general information on what genetic disease entails and a clarification of the very misunderstood concepts such as inheritance, carrier status, and the chance nature of chromosomal abnormalities. Not only must the guilt associated with genetic disease be dispelled, but also people should be encouraged to discuss the ramifications of genetic disease as to responsibility to the affected persons, the family, as well as society. This is best accomplished through counseling or public interest program

In addition to educating the public on genetic disease in general, specific education programs must be included in all legislation relating to screening, counseling, or other programs designed to reduce the incidence of genetic disease. The public, and especially certain high-risk groups, must not only be made aware of the services available but also must be educated on the scope and limitations of each program. While this education should be designed so as not to lead to false hopes, the importance of specific intervention techniques should be presented in a balanced and objective manner so that informed consent is possible. While information concerning these programs must be intensive among those at risk, the more inclusive attempt to educate should be as broad-based as possible. The disastrous attempts at sickle-cell screening in some states where carriers were led to believe they suffered from the disease should serve as examples of how not to conduct a program. The well-organized and community-controlled education procedures used in the Washington, D. C., Tay-Sachs screening effort, conversely, should serve as a model of a successful program.

Societal Responsibility and Genetic Disease: Conclusions

In analyzing any of the human genetic intervention programs discussed here, as well as those such as eugenic sterilization, limits on childbearing age, and sperm banks, it is crucial to examine the responsibility of society to a wide variety of interests. Recent history has been one of wide fluctuations as to where primary concern is directed. Early eugenic attempts had wide support, as is apparent in Justice Holmes's affirmation in *Buck vs. Bell* of the right of the state to sterilize in the name of the public welfare in order to prevent society from being "swamped with incompetence."

Recently, emphasis has shifted to the right of parents, despite their competence, to bear children. Procreation is perceived by many as an interest so fundamental that society has no right to intervene. Both of these absolutist positions neglect the primary responsibility to those who would be affected by genetic disease if born and to future generations. Too often the policy choice is based solely on either: (1) the rights of the parents to reproduce at will, or (2) certain inherent powers of the state to protect the health of its citizens.

It must be noted that all rights are relative to other rights, not absolute. While society cannot indiscriminately violate the rights of parents and their reproductive autonomy, it is reasonable that society also has a responsibility to those directly affected by the broad range of genetic diseases. It is grossly unfair to concentrate on the rights of today's parents without accounting for the concerns of those potentially or actually afflicted or to some broader societal needs. Certainly, the fundamental right to bear children, although at the core of any genuine democratic society, has limits. Whether the imposition of any specific genetic intervention on this right is justifiable or not within the broader context of social responsibility is dependent on the merits of the particular program.

While no screening policy should be accepted without a careful and intensive analysis of its implications for individual rights and a weighing of the responsibilities to the many concerned parties, no screening or diagnostic decision can be rejected out of hand solely on the basis of the rights of any one of the parties. Although it would be folly to accept at face value the techniques for genetic intervention and impose them without a critical and rigorous examination of their implications, it is just as foolish to dismiss them in the absence of such analysis. Genetic disease is real and the state of technology is advancing the potential benefits of intervention at a rapid pace. Now is the time to examine the advantages and disadvantages and weigh societal responsibility in the reduction of genetic disease through the various means available.

Notes

1. For a good discussion of this, see Douglas (1976).

2. Feinberg (1974), in examining environmental protection, sees concern for our children as a matter of love, while that for our remote descendants is a matter of justice, of respect for their rights as humans. In his view, we do not have a right to deprive future generations of the necessary conditions for life as we know it. He acknowledges that justice implies restrictions on certain freedoms of those now living. We have the duty to refrain from polluting the environment, exhausting nonrenewable resources, and procreating an unmanageably large number of people.

3. Levy (1973:72-73) makes a distinction between screening and diagnosis. The term "screening" here will at times be used to refer to both collectively.

4. For an excellent summary of the technological state of prenatal diagnosis, see Nadler (1972).

5. Initial success in isolating male fetal cells from a blood sample of the mother was achieved at Stanford Medical Center in August, 1977.

6. In every case, the female will be a carrier of the disease but will not be affected.

7. See Karp (1976) and Schelling (1978) for discussions of this. Schelling in chapter 6 extends this to other selection characteristics as well as sex.

8. Bessman and Swazey (1971) offer a critical account of early PKU legislation and argue that it was premature. Also see Murray (1972:11-13).

9. For a complete description of particular provisions of each state's legislation see NAS (1975:98-102). The District of Columbia has since dropped its PKU program after several years of screening without identifying a single case since PKU is rare among blacks, which constitute 70 percent of its population.

10. A more complete description of the Washington screening program is found in Kaback and O'Brien (1973).

11. Also see Lappé (1975) and Heim (1973) for further discussion and additional views.

10

Intelligence Theory and the Race-IQ Controversy

JOHN G. BORKOWSKI

There are three major issues in the race-IQ controversy. Are there group differences in intelligence between blacks and whites? If so, are these differences due to genetic or to environmental factors? Are differences modifiable? While these questions remain central to the race-IQ controversy, they cannot be addressed adequately until related issues are more clearly delineated (cf. Ehrlich and Feldman, 1977; Vernon, 1979). For instance, what is race? What is heritability? How does heritability affect intelligence? What factors in the environment contribute to changes in intelligence? How might heredity and environment interact?

In spite of the importance of these secondary issues for sharpening the focus of the main controversy, a more critical question, often overlooked in the debate, requires clarification and elaboration. This question concerns the nature and measurement of intelligence itself.

While the traditional tests of IQ, such as Binet's and Wechs-

The writing of this essay was supported in part by a grant from the National Institute of Education, Department of Health, Education, and Welfare (No. NIE-G-79-0134). The opinions expressed herein do not necessarily reflect the position or policy of the National Institute of Education, and no official endorsement by the National Institute of Education should be inferred.

ler's, continue to be used by professional psychologists in forms essentially unmodified for decades, the concept of intelligence upon which these tests were originally based is undergoing intense empirical scrutiny and theoretical modification. Works by Resnick (1976), Horn (1978), Campione and Brown (1978), and Sternberg (1979) are but a few of many recent papers and books reflecting the resurgence of interest among psychologists about a monumental question: What is intelligence?

In spite of renewed interest in revising intelligence theory, serious analyses of the race-IQ controversy continue to employ a data base that rests almost exclusively on traditional tests of IQ (cf. Loehlin, Lindzey, and Spuhler, 1975; Ehrlich and Feldman, 1977). These traditional IQ tests (e.g., the Stanford-Binet and Wechsler Intelligence Scale for Children) rely on conceptualizations about intelligence formed primarily by Binet, Spearman, Thurstone, Burt, and Wechsler, perspectives essentially unchanged in structure and focus for eight decades. In short, standard IQ tests, and their underlying theories, do not reflect the intricacies of current perspectives about memory, cognition, and other aspects of information processing. Rich, new insights are emerging about the nature of intelligence to guide the modification and reconstruction of standardized IQ tests (cf. Resnick, 1976; Sternberg and Detterman, 1979).

The state of our knowledge relevant to race-IQ questions is confused and controversial, in large part, because of imprecise and restricted theories about the nature of intelligence. Additional scientific analysis, in terms of new theory and supporting research, may yield a more accurate and constructive resolution of the central questions about racial differences in intelligence. The purpose of this essay is to revisit race-IQ questions from a different theoretical perspective, the Campione-Brown (1978) model.[1] The essay analyzes existing data on race and intelligence in light of the cognitive development model of Campione-Brown and speculates about the origins of observed differences. Specifically, do blacks and whites differ in specific structures or processes necessary for efficient problem-solving, and how might differences be related to environmental influences? New theoretical interpretations, implications for teaching, and research directions flow from the reanalysis of existing data.

Changing Concepts about Intelligence

Although there is no single agreed-upon concept of intelligence (Horn, 1978), most widely used measures of IQ such as the Binet and Wechsler tests are viewed by psychologists as indicants of the capacity to adapt, to abstract, and to invent. There is a certain paradox in this view of intelligence. On the one hand, the concept represents potential for problem-solving in the face of challenge. On the other hand, an IQ test must reflect skills that are capable of being reliably measured, such as vocabulary and information. Items in IQ tests often appear only tangentially related to actual problem-solving and remote from the act of invention. In short, an ever-widening gap separates the concept of intelligence as potential for adaptation, problem-solving, and invention from its operational measurement in the traditional tests of IQ.

The Traditional View of Intelligence

The construction of individual intelligence tests reflects two lines of thought about the nature of intelligence. Spearman and Binet suggested that our diverse and often uneven abilities were manifestations of a single underlying function (labeled *g* by Spearman). The central idea is that only one basic component of intelligence produces our distinct abilities. Later theories (Burt, 1949; Cattell, 1971; Horn, 1968; Jensen, 1979) can be seen to retain the *g* factor; they differ mainly in terms of the nature and specificity of subfunctions that contribute to *g*.

A second type of theory emphasizes independent primary abilities. L. L. Thurstone (1947) identified nine primary mental abilities that reflect the diversity of human intelligence. Guilford (1967, 1979) has developed a structure of intellect model in which abilities, and the tasks that measure them, can be organized around five kinds of operations (evaluation, convergent production, memory, divergent production, cognition), four kinds of contents (semantic, symbolic, figural, behavioral), and six kinds of products (units, classes, relations, systems, transformations, implications). The interaction of these abilities defines intelligent behaviors.

Both theoretical concepts—intelligence as *g* and as primary mental abilities—enter into the construction and scoring of our contemporary IQ tests. In the Wechsler test, ten subtests reflect different, although sometimes overlapping, abilities (such as information, short-term memory, etc.). In the Binet test, individual questions are designed to measure distinct abilities (e.g., verbal skill, memory, reasoning, quantitative skill, etc.). In both tests, commonality in performance among subtests or among diverse items reflects general intelligence. It is from this theoretical perspective of intelligence, defined by the *g* factor plus separate primary abilities, that the race-IQ issue has been analyzed, researched, and interpreted (cf. Loehlin, Lindzey, and Spuhler, 1975).

Using the traditional definition of intelligence as a framework various questions have been asked about race and IQ. Are group IQ scores identical for black and white individuals (Jensen, 1969)? Is heritability in IQ similar across racial groups (Jensen, 1969; Ehrlich and Feldman, 1977)? Does cross-racial adoption change IQ scores (Scarr and Winberg, 1976)? Do some abilities differentiate racial groups more than others (Lesser, Fifer, and Clark, 1965; Wysocki and Wysocki, 1969)? The last question—the interaction of specific abilities and racial groups—has been asked less frequently than other questions. Ironically, it is probably the most theoretically interesting and educationally relevant.

In 1975, Loehlin, Lindzey, and Spuhler urged social science researchers to continue their exploration of race-IQ issues, but to shift the thrust of their research directions and to focus on comparisons of racial-ethnic groups along a number of discrete abilities. They suggested that causal links between abilities and specific genetic and environmental factors be determined. The present approach follows these suggestions. That is, the goal of this essay is to offer a revised model of intelligence in order to redirect theoretical interest and research attention about race-IQ issues from an analysis of overall Wechsler and Binet test scores to more specific assessments of structure and process components of intelligence. An information-processing-based model of intelligence and a knowledge of the modifi-

ability or resiliency of its components likely will lead to more
accurate knowledge and educationally relevant perspectives
about race and intelligence. The remaining sections of this
paper sketch a cognitive development model of intelligence,
compare racial groups in terms of the model's components,
and discuss policy implications.

A Cognitive Development Model of Intelligence

Common to most traditional tests of IQ is the failure to
distinguish between two fundamental types of skills that differentiate retarded, normal, and gifted individuals: (1) skills closely linked to the sensory abilities of the organism, deeply rooted
in central nervous system integrity, and essential for primitive
cognitive operations such as perception and immediate memory;
and (2) skills related directly to experience with or training in
complex problem-solving, that are highly modifiable, and that
distinguish creative, adaptive learning from rote non-strategic
learning.

Emerging from their research on memory processes in the
mentally retarded, Campione and Brown (1978) have provided
a fresh perspective on the nature of intelligence. Their theory
postulates two hierarchical levels: (1) an architectural system
that is characterized by perceptual efficiency in encoding-
decoding information; and (2) an executive system[2] that ini-
tiates and regulates retrieval of knowledge from long-term
memory, Piagetian schemes of thought, rehearsal strategies or
control processes, and metacognitive states. The complex and
intricate interaction of the architectural and executive systems
is necessary for decision making and problem-solving with difficult tasks. Recent research with normal children (Ornstein,
1978) and mentally retarded children (Ellis, 1979) provides
the beginnings of the empirical foundation for the theory.
According to the Campione-Brown model, intellectual performance in the face of complexity is determined by the components of the architectural system, the executive system, and
their interaction.[3]

Architectural-Executive Theory: Assessment Procedures

At present there is no formal assessment battery that will measure intelligence in terms of the functioning of architectural and executive systems. No existing tests accompany the Campione-Brown (1978) prescriptions. However, we can examine a number of isolated studies that assess various aspects of perceptual efficiency. Measures of speed of encoding, rate of memory search, memory span, and reaction time to simple and complex tasks can be considered as indicants of perceptual efficiency. By and large they measure speed and efficiency in the selection, storage, and retrieval of information over brief intervals of time.

In the existing psychological literature, a number of studies compare racial groups on measures of perceptual efficiency. Studies of digit and word span, recall in primary memory, and reaction time have contrasted the performance of blacks and whites in terms of their processing efficiency at the system's most fundamental level, its perceptual hardware or architecture. In future research, measures such as stimulus persistence in backward masking paradigms (Birren, 1974), inspection time (Brand, 1980), decoding speed in Posner-type judgment tasks (Posner and Boies, 1971), and scanning speed in the Sternberg (1966) task should also be compared and evaluated as potential indices of perceptual efficiency. The present analysis of racial differences in perceptual efficiency represents only a beginning step in the extension of contemporary views of intelligence to race-IQ issues.

It should be noted that some of the components of the executive system, such as knowledge, schema, and metacognition, are complementary, overlapping hypothetical constructs. Although each appears capable of independent operational specification, research may reveal that the preliminary theoretical rationale for attempting to separate the components is invalid. For instance, schemata, knowledge, and metacognition may be parts of a coherent theoretical network, reflective of the same processes or structures. For the moment, however, it seems prudent to consider the overlapping components of the executive system as independent concepts, in terms of their theoretical bases and assessment procedures.

Three components of the executive system—knowledge, control processes, and metacognitions—can be directly measured. The measurement of Piagetian schemes—as reflected in developmental changes in cognitive states—awaits additional research such as an extension of the scale of sensorimotor development developed by Uzgiris and Hunt (1974). Knowledge can be measured by borrowing the information, vocabulary, and arithmetic subtests from the Wechsler tests or by more domain-specific tests. Control processes can be measured by observing strategic behaviors independent of the performance that they sometimes determine. For instance, clustering in free recall, interacting imagery in paired-associate learning, and changing pause-time patterns in serial learning are examples of behaviors that reflect control processing. Finally, metacognition—knowledge about when and how to be strategic—can be tested by using interview questions similar to those formed by Kreutzer, Leonard, and Flavell (1975) to measure children's awareness of how their memory system operates. Gilewski, Zelinski, and Thompson (1978) have developed a rating scale for adults in which questions are asked about frequency of memory problems and effort used in remembering. As existing research on the executive system is reviewed in the next section, it should be noted that most studies contrasting racial groups have concentrated on tests of knowledge states. Future research needs to focus more on control processes and metacognitive states in order to examine racial similarities and differences in the operation and modification of the executive system.

Comparisons with Jensen's Levels I and II Abilities

Is a fresh analysis of old data on a socially explosive issue justifiable? Although there are many facets to this question, there is a reasonable chance that new insights and understanding will be gained on why black children often score lower on IQ tests than white children. The potential for increased understanding can be gleaned by comparing the orientation of Campione-Brown with other positions on the nature of intelligence. In 1969, Arthur Jensen proposed a two-dimensional theory that tried to account for individual differences in intelligence. Level I, referred to as associative ability, is measured by tests such as digit span, serial and paired-associate learning, free recall, and trial-and-error learning. Level II, or conceptual ability, involves self-initiated elaborations and transformations of incoming information; its measurement is found in problem-solving and concept-learning tasks as well as in tests with high g loadings such as the Raven's Progressive Matrices. Level I is assumed necessary but not sufficient for Level II functioning, which is akin to Spearman's g factor.

Lower-class children, black and white, were found by Jensen to perform similar to middle-class children on tests of digit span and paired-associate learning but differed widely on tests of concept learning. These facts led to the conclusion that Level I skills are distributed about evenly across social classes and races, while Level II skills are more evident in middle- and upper-class whites. Hence, differences between black and white children emerge as Level II ability becomes more involved in a task's solution.

There are several criticisms of Jensen's theory that are avoided in the Campione-Brown theory. Jensen's Level I, associative ability, is subject to the same criticism that cognitive theory brings to the evaluation of learning theory (Brewer, 1974). Is the concept of associations useful and accurate in describing how children actually solve learning problems, even the so-called rote tasks such as paired-associate, free recall, and serial learning? Children beyond the age of five often mentally transform associative tasks so that strategies are available to aid performance (Borkowski and Cavanaugh, 1979; Ornstein, 1978) The presence of strategic behaviors makes difficult the measurement of "pure" associations. Since young children sometimes use rather complicated strategies to learn "associative" tasks (Neimark, 1976), elevating a Level I task to a Level II task, the notion of association is in disfavor among many cognitive developmental psychologists as an appropriate metaphor to describe learning (cf. Rohwer, 1971).

As a substitute for Level I associative ability, the Campione-

Brown conception of intelligence offers the construct, perceptual efficiency. Efficiency refers to the speed with which information enters the information-processing system and the speed with which it is retrieved when needed. Rehearsal rate (Salthouse, 1979), inspection time (Brand, 1980), span ability (Bachelder and Denny, 1977), reaction time (Jensen, 1979), and memory search (Sternberg, 1966) are examples of ways to test perceptual efficiency.

Jensen (1969) states that Level II abilities should be differentially distributed across social classes. On this point an analysis from the perspective of the Campione-Brown theory seems to agree. However, the reason for the differential distribution of Level II abilities are quite different. Jensen argued for different inherited neural structures underlying Level II skills, while advocates of the Campione-Brown model stress the importance of child-rearing practices and quality of education as factors influencing executive skills. These Level II-type executive skills include knowledge, strategies for obtaining new knowledge, and awareness of when, where, and how to use strategies.

Environmental factors are postulated as important determinants of executive functioning for two reasons: (1) Strategic behaviors in retarded individuals are highly modifiable. For instance, Kendall, Borkowski, and Cavanaugh (1980) showed that retarded children can acquire, with little effort, a rather complex self-interrogation strategy, use it to learn a paired-associate task, and generalize parts of the strategy to a different but related task. (2) The quality of education seems to influence the type and efficiency of executive processing. Recently, Stevenson, Parker, Wilkinson, Bonnevaux, and Gonzalez (1978) reported differences in the ways schooled and unschooled Peruvian children process information; cognitive skills result more from educational opportunities than from social class. From the Campione-Brown perspective, we look to the quality of stimulation in the environment as important in determining the operation of the executive system. This orientation, if supportable, might yield a revised understanding of the origins of race-IQ differences.

What emerges in the Jensen system are associative and conceptual levels, both highly genetically linked. In the Campione-Brown theory, perceptual efficiency, defined as speed of encoding/decoding information, and the executive system are the central concepts. The latter component is known to be highly susceptible to environmental modification for all children not grossly limited by problems in perceptual efficiency.

Racial Differences in Intelligence

There are only a few studies in which more than one component of the architectural and executive systems have been measured in the same individuals, either black or white. Thus the issues under consideration here can only be approached by analyzing how racial groups respond to tests of the architectural and executive systems in a series of somewhat unrelated studies. Such an approach has pitfalls. Are samples equivalent across studies? Are seemingly identical tests actually administered in identical ways? Are potential confounding factors, such as socioeconomic status (SES), controlled? While methodological problems make firm conclusions impossible, it is instructive to revisit existing data to determine if alternative interpretations or hypotheses seem reasonable in light of new concepts about the nature of intelligence.

We begin our review of the literature with a prediction drawn from the Campione-Brown model: there is an interaction between racial groups on tests of the architectural and executive systems. The more specific hypotheses related to this interaction are straightforward and testable. On tests of perceptual efficiency black and white people should perform similarly. In contrast, sizable differences should emerge on tests of knowledge, control processes, and metacognitive states (e.g., awareness about when, where, and how to be strategic).

Racial Differences in Perceptual Efficiency

In 1923, Clark found that black and white adolescents had equivalent digit spans, defined by the maximum number of digits that are recalled immediately after their presentations, although the latter had a higher overall IQ. Since then, hundreds of studies have tested black and white children on various IQ tests, many of which have assessed digit span (cf. Shuey, 1966). For the most part, differences due to race are minimal when span ability is examined in test situations that preclude the use of grouping strategies, such as occurs when words are presented at fast rates of presentation.

Young and Pitts (1951), Caldwell and Smith (1968), and Wysocki and Wysocki (1969) compared large numbers of black individuals on the various subtests of the Wechsler tests. In general, digit span ranked high among the ten subtests in terms of relative performance. In a more complex study using a factor analytic approach, Sitkei and Myers (1969) found significant differences on tests of vocabulary and verbal comprehension in four-year-old children due to SES and to race, whereas no difference was found on a memory span test. More recently, Hall and Kaye (1977) tested 600 boys (ages five, six, seven, and eight) on tests of digit span and general intelligence as indexed by Peabody Picture Vocabulary and Raven Matrices tests. There were sizable effects of age, social class, and race on the two measures of general intelligence. However, only age effects were found on tests of digit span. There were no span differences between black and white children in lower- or middle-class groups. The latter finding is particularly important to the prediction of no racial differences on measures of perceptual efficiency. The findings of Hall and Kaye (1977) are consistent with data provided by Jensen and Figueroa (1975) and Grimmett (1975). These studies showed that forward digit span is unrelated to race when SES matched groups are compared.

Supporting evidence for the contention that there are no racial differences in perceptual efficiency has been provided by Ellis (1970) and Salzinger, Salzinger, and Hobson (1967). Ellis compared thirty-four children (thirty-one of whom were black) in a Head Start program with twenty white kindergarten children on a six-position memory probe task. No differences in primary memory—the recall of the most recently presented digit—were observed. Salzinger et al. (1967) compared middle-class white children with lower-class black children on a sentence imitation task. The white children showed greater accuracy in imitating simple declarative sentences, presumably because of more exposure to correct syntax in their language environments. However, when required to imitate reversed sentences ("Eggs eats mouse the"), the black children performed slightly better than the white children. The interpretation here is that reversed imitation is similar to a word span test, a measure of perceptual efficiency, and is relatively independent of language proficiency. In both the Ellis and Salzinger studies it should be noted that social class and race were confounded—with the direction of the confounding providing an advantage to the white groups. Nevertheless, no differences due to race were found on tests of perceptual efficiency. In sum, all of the studies reviewed to this point suggest a common theme: racial differences are absent on measures of perceptual efficiency, the key component of the architectural system.

In contrast to this interpretation, Jensen (1979) has found slope differences in reaction times (RT) to simple and complex tasks. As task complexity increased, differences in RT slopes were observed for the black and white male vocational school freshmen matched on aptitude test scores. If differences such as these prove reliable across a wide age range, and if RT slope differences prove valid in terms of their connections with other measures of perceptual efficiency, then the RT data of Jensen (1979) will stand in contradiction to the thesis that racial differences are absent in tests of perceptual efficiency. Obviously, further tests of RT differences between racial groups are needed, as well as comparisons on other possible measures of efficiency such as inspection time, stimulus persistence, scanning speed, rehearsal rate, and decoding speed. At this early stage, however, it seems reasonable to suggest that racial differences appear to be minimal on several tests of perceptual efficiency. The fifteen-point difference in IQ found between blacks and whites can only be attributed in small part, perhaps not at all, to racial dissimilarities in perceptual efficiency.

Racial Differences in Executive Functioning

Knowledge. A sizable body of data implicates knowledge and information as states that differentiate the executive systems of black and white children. Furthermore, racial differences in knowledge appear to increase with age (Baughman and Dahlstrom, 1968; Hall and Kaye, 1977). Young and Bright (1954), Caldwell (1954), and Stodolski and Lesser (1967) noted poorer performance by black children on the vocabulary test of the WISC. In an extensive study of black and white children in the rural South, Baughman and Dahlstrom (1968) found sizable racial differences in word meaning, paragraph meaning, arithmetic reasoning and computation, and language usage. Composite scores on the Stanford Achievement Test Battery are shown in Figure 1. It can be seen that differences in amount of knowledge about words and numbers increased with age and were apparently linked to race. It should be emphasized that all of the studies cited in this section on knowledge differences failed to separate the SES factor from racial comparisons. Hence, differences in vocabulary and more general knowledge may be due to race, SES, or other factors such as quality of education.

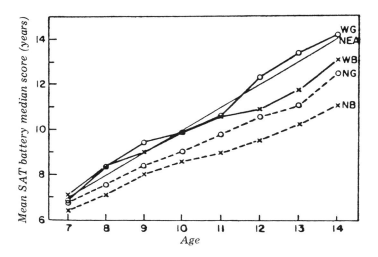

Figure 1. Stanford Achievement battery composite scores as a function of sex (B and G) and race (N and W) for number and words. Reprinted by permission from E. E. Baughman and W. G. Dahlstrom, *Negro and White Children: A Psychological Study in the Rural South* (New York: Academic Press, 1968).

In 1977, Hall and Kaye controlled for SES and tested large samples of six-, seven-, and eight-year-old white and black children on a variety of learning, memory, and achievement tasks. Of relevance is the data shown in Figure 2, comparing racial-

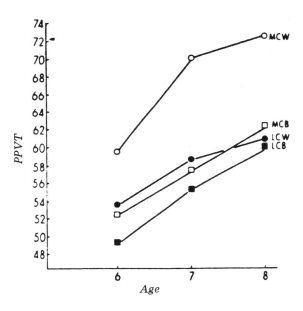

Figure 2. Mean scores on the Peabody Picture Vocabulary Test as a function of race and SES. Reprinted by permission from V. C. Hall and D. B. Kaye, "Pattern of Early Cognitive Development Among Boys in Four Subcultural Groups," *Journal of Educational Psychology* 69 (1977): 66-87.

SES groups on the Peabody Picture Vocabulary Test. What is striking is the race by social class interaction. Further analyses showed the dependency of racial comparisons on level of SES. That is, superior performance of the white children occurred only in the middle class. Similar results were observed in comparing the children's performances in the Raven Coloured Progressive Matrix. It should be emphasized that on a test of perceptual efficiency (digit span), black, older, and middle-class children performed better than the other groups. These results suggest that on a measure of perceptual efficiency, racial differences were minimal, whereas on tests of the executive system (knowledge) they were apparent in middle-class children. In a somewhat similar study, Sitkei and Meyers (1969) also found racial differences in vocabulary and verbal comprehension tests with SES controlled. However, in the Sitkei and Meyers study, differences were present in both lower- and middle-class samples of four-year-old children.

In summary, there is a considerable body of evidence suggesting that racial differences exist in various tests of knowledge states, especially vocabulary, verbal comprehension, and number skills. Differences are observed when black and white children are compared without consideration of SES factors. However, racial differences in verbal knowledge remain when groups are matched on SES and seem especially prevalent in middle-class comparisons.

Control processes. Few studies have directly compared black and white children, matched on SES, on both control process and learning performance measures. The reason that control processes, such as strategic skills, are important in defining intelligence is that they often lead to high levels of performance accuracy. Performance and the strategies used to produce it can be measured independently in many learning tasks (cf. Belmont and Butterfield, 1977). The assessment of important strategic behaviors in multiple tasks defines an individual's level and style of control processing. At a higher cognitive level, changes in strategic behavior in response to a change in task demands reflects executive functioning (Butterfield and Belmont, 1977). Appropriate changes in executive functioning in the face of changing tasks are the hallmarks of creative, inventive forms of intelligence (Belmont, 1978). Essentially, the rationale for measuring control processes is that they often determine complex, adaptive behaviors. The next section compares racial groups on a single measure of control processing. More extensive racial comparisons of control processes and executive functioning await future research.

Jensen and Fredericksen (1973) used a free recall task to measure correct responding as well as associated clustering strategies in second- and fourth-grade black and white children on categorized and uncategorized lists. No significant differences were found between groups at either grade for uncategorized lists; it should be noted that unrelated words are not easily clustered into manageable units. Similarly, recall on the categorized list for the younger black and white children was equivalent. In contrast, older white children showed significantly more clustering and better recall than older black children on the categorized list. Since SES was not controlled in this study, it is unclear whether factors related to race, SES, or both were associated with differences in strategic behavior found on the categorized list for the fourth graders. Whatever the cause, older black children showed less spontaneous clustering—a measure of control processing—than white children. Younger children did not show differences in clustering.

There is an important point to make about control processes: they are easily modified. While we are uncertain of the environmental factors that give rise to possible racial differences in control processes, differences can probably be reduced or eliminated through strategy training. For instance, Scrofani (1972), Grimmet (1975), and Reichart and Borkowski (1978) have shown that clustering strategies, such as those measured by Jensen and Fredericksen (1973), can be quickly acquired during pretraining and generally will improve recall on transfer tests. A reasonable working hypothesis is that the acquisition, maintenance, and generalization of strategic skills are not limited by race or SES. From this premise, it can be argued that whatever differences exist in control processes among racial groups can be altered through appropriate remedial training (cf. Borkowski and Cavanaugh, 1979).

Metacognition. A considerable amount of research effort and theory development has been directed toward clarifying the concept of metacognition—the awareness of when, where, why, and how to be strategic in solving problems (cf. Brown, 1978; Flavell, 1978; Sternberg, 1979). Since this work is recent it should not be surprising that there are no data comparing black and white children on metacognitive tasks such as recall readiness, span estimation, or judgments about the efficiency of memory processes or strategies. Nevertheless, several hints in the literature suggest that black and white children may have

different degrees of awareness about the value of strategic behavior as aids to successful problem-solving. For instance,
Jensen and Fredericksen (1973) provided children a word list
grouped by categories (all members of one category were shown,
then all the members of the next category, and so on), rather
than a randomly presented word list. Racial differences in recall
and clustering were reduced in the "grouped" condition, due
mostly to improved performance for the black children on
later trials. Loehlin et al. (1975) interpreted this finding in
terms of a differential readiness to employ a classification strategy. Differential readiness is an aspect of metacognition—knowledge about when and how to use a strategy.

Hall and Kaye (1977) found that white children improved
more on a free recall task from trial one to trial two than did
the black children of the same SES. This finding is perhaps indicative of a greater metacognitive readiness on the part of the
white children to use a clustering strategy. Obviously, more
research is needed to assess whether black and white children
differ in their states of awareness about when and how to use
strategic behaviors as aids to problem-solving. As with control
processes, most racial differences in metacognitive states can
probably be reduced through carefully crafted, intensive educational programs.

Summary and extension. The major interpretations of the
data presented in the preceding section are as follows: (1) When
viewed from the perspective of the Campione-Brown theory of
intelligence, racial differences across a wide age range are absent,
or at least minimal, on tests of perceptual efficiency. (2) Sizable
differences emerge on one subcomponent of the executive system, knowledge states (e.g., vocabulary). (3) Limited evidence
points to racial differences in the use of control processes. (4) No
firm conclusions can be drawn about racial differences in metacognitive states, although it is hypothesized that important differences exist in this component of the executive system.

What is the course of intellectual development from birth
through adolescence for black and white children? One view is
that racial differences in intellectual abilities are nonsignificant
during the first few years of life (cf. Loehlin et al., 1975), but
diverge beyond age four and continue to increase through the
school years (Baughman and Dahlstrom, 1968). However, this

rather standard conclusion is probably imprecise as well as misleading since it is based on results from traditional models of intelligence. When reanalyzed from the Campione-Brown theory, there appears to be little support for the notion that blacks and whites differ on measures of perceptual efficiency at any age. In contrast, differences in executive functioning, especially in general knowledge states (and perhaps control processes and metacognitive knowledge), should increase as children proceed to preschool and through grade school. This hypothesized age-by-race interaction in components of the executive system should be highly dependent on a third factor, quality of training in cognitive processing. The next section addresses possible reasons for the age-by-race interaction in executive skills.

Reanalysis of the Race-IQ Issues: Policy Implications

Factors That Produce Executive Skills

The slowed development of the executive system in many black children and low SES children[4] of all races may be attributable to two factors: (1) less emphasis in the home on language interactions and language-related conceptual skills; (2) less educational enrichment in the school leaving children unprepared, in terms of available strategies and plans, to confront complex learning and problem-solving situations, especially in the advanced grades.

Hess and Shipman (1965) documented the importance of family structure in shaping effective communication in the young child and in developing cognitive styles appropriate for problem-solving. Levenstein (1975) showed that when mothers are taught to verbally interact with their children in a systematic way, the children's WISC IQs increased by about fifteen points following two years of intervention. Vocabulary, math, and social skills also increased significantly as a result of more active, structured verbal interactions between mother and child. As Levenstein (1975) states:

> Many mothers, particularly those of the college-educated, middle income group, carry on such "intervention" spontaneously, as part of a now well known "hidden curriculum" of the family. Increased cogni-

tive development results for the child. Schaefer, indeed, suggested that early, basic informal home education be given a name, "Ur-Education," and be recognized as a legitimate supplement to conventional academic education (1970). Implicit in the concept is the inseparability of the mother-child emotional interchange from the cognitive interaction, and intermingling of cognitive stimulation and deepening of the attachment between the mother and child.

Levenstein's results are in accord with the findings of the Milwaukee project (Garber and Herber, 1973). Children at risk for mental retardation were placed in early childhood stimulation programs within the first six months of life. In addition, their mothers were placed in a vocational-personal rehabilitation program. Dramatic gains in IQ, language skills, and strategic behaviors were shown by the children given extensive, early training in cognitive skills. The IQ gain for children at age five was around twenty-five points. Collectively, the findings of Hess and Shipman (1965), Levenstein (1975), Garber and Herber (1973), and others emphasize the importance of early environmental stimulation on later intellectual development, especially for advances in knowledge states and cognitive strategies.

If knowledge about the world, language fluency, computational skills, problem-solving strategies, and an awareness of cognitive states and processes are to fully develop, then early stimulation in home and preschool environments seems essential. Environment enrichment likely combines with perceptual efficiency to dictate the developmental pattern for components of the executive system. A deficit in either perceptual efficiency or cognitive stimulation will likely prevent (in the case of an efficiency deficit) or delay (in the case of slowed executive development) overall intellectual growth.

Intelligence Theory and Educational Goals

What should education be about? It is difficult to argue against the traditional goals of reading, writing, and arithmetic.

But what seems lacking from many educational programs are equally important goals: teaching children how to read, how to write, and how to solve problems systematically and strategically. What is implied in the present analysis is that major objectives during the first few years of school are training control processes and teaching ways to learn prior to telling what to learn (cf. O'Neil, 1978).

Jerry Lucas's (1978) popular manual on making learning efficient and fun for children, *Ready, Set, Remember*, provides a good example of how school can be made easier when strategies are taught before content. Children instructed on how to use imagery as a control process find learning fun, fast, and durable. In a similar vein, Levin (1977) suggests that strategy training—teaching new ways to learn information—should occupy a part of each child's school day. From this perspective, a major objective of education is the elevation of intelligence through systematic training of control processes and an increase in metacognition, the appreciation for the value of plans and their potential applications. After appropriate control processes are mastered, domain-specific knowledge acquisition and general problem-solving will likely be enhanced. One result is that the educational process might become more enjoyable, meaningful, and useful to a wide range of children.

Schools for middle and upper SES children, many of whom are white, have tended to focus not only on content but also on the development of strategy-based learning and the integration of background knowledge with special strategies to solve new problems. In contrast, the quality of education provided for blacks has traditionally been inferior to that provided for whites (Weinberg, 1977). One consequence of educational inequities is to further separate racial groups in terms of opportunities to acquire new executive skills, techniques for approaching academic tasks systematically and confidently. It is likely that educational inequity has not only led to immature learning strategies for many black children, but also to less general information and knowledge in their memories. Our hypothesis is that the more efficient the learning methods used during the child's early years, the richer and fuller their knowledge state in later years.

The use of strategies to solve problems and to acquire new facts promotes metacognitive growth. Active, strategic learners have a greater awareness of the need to be systematic and flexible in solving problems. They come to invent unique strategies in the face of novel, challenging problems. The act of invention is a sign of more advanced levels of human intelligence (cf. Belmont, 1978). Invention most likely emerges when there is coordination among knowledge states, control processes, and metacognition states. It is the job of education to produce such integration efficiently and smoothly. As cognitive and educational psychology learn how to coordinate their research efforts, methods for achieving this coordination for each child become more realistic. A failure of the system to respond to that challenge—as has occurred in the past for many economically disadvantaged children, both black and white—will continue to result in lower levels of intellectual functioning in components of the executive system.

If the present analysis is correct, then racial differences in intellectual functioning that frequently appear on tests of executive skills will disappear or be greatly reduced, provided that cognitive intervention in the home and in the school gives more opportunity for black children to develop knowledge, control processes, and metacognitive skills. With rapid advances in research and theory aimed at linking intelligence to education, future questions about race and IQ might well become irrelevant. More important issues, beneficial for all children, would emerge: Given a child's present functioning in terms of perceptual efficiency and executive skills, what can be done to optimize performance on intellectual tasks effectively and efficiently?

Summary

When viewed from a cognitively based theory of intelligence, much of the fifteen-point spread in IQ scores separating black and white children on Binet or WISC tests might be attributable to the functioning of the executive system as it influences responses to complex problems. Research is needed to test this prediction. Of special importance for theory development is the nature of the interaction between the architectural and executive systems, between perceptual efficiency and the child's knowledge base, strategic skills, and sense of awareness about their possible integration while confronting novel problems. Mapping this interaction in terms of new theory seems necessary for understanding how children solve complex conceptual problems.

The present analysis suggests that racial differences in skills of the executive system exist primarily insofar as environmental stimulation, such as the development of study strategies, favors the members of one racial group over another. In our society, many low SES children do not receive adequate training of executive skills. The educational advantages—in the home, in preschool, and in the early grades—that middle- and upper-class whites have traditionally enjoyed are probably responsible for many differences on IQ questions that tap components of the executive system. What should be emphasized is that racial differences may be minimal or absent in architectural components of intelligence.

Perhaps the main implication of this revised look at the race-IQ controversy, in light of the Campione-Brown theory of intelligence, is the following: Rehearsal strategies, decision-making routines, and metacognitive processes can be radically altered in young children through systematic, intensive training based on techniques used in the instructional approach to cognitive development (Belmont and Butterfield, 1977) or in cognitive behavior modification (Meichenbaum and Asarnow, 1978). Improvements in performance expected from cognitive interventions, aimed at reshaping the executive system, have no known racial limitations. In fact, elevating information-processing skills for black children might be expected to produce greater improvement in performance than for white children.

Notes

1. It should be noted that other contemporary views of intelligence, such as Sternberg's (1979) or Pellegrino and Glaser's (1979), provide rich perspectives for a reanalysis of the race-IQ controversy. My preference for beginning this reanalysis with the Campione-Brown model is based on the following observations: (1) a focus on the architectural

system that is not featured in Sternberg's componential model and (2) a clearer specification of the nature of the executive-control system than in the Pellegrino-Glaser model.

2. This term is introduced to designate the higher level system originally described by Campione and Brown to reflect components of long-term memory.

3. A more complete treatment of the components of the architectural and executive systems can be found in Borkowski (1980).

4. It should be pointed out that the preceding discussion of black-white differences on tests of intelligence attempted, in a very general way, to separate SES factors from racial comparisons. In the population as a whole, however, race and SES are confounded. Hence, most of the implications in this section apply not only to racial groups but also to low SES children and possibly to some children in the middle SES range, depending on the quality of stimulation in their homes and educational programming in their schools.

11
Summary and Conclusions

MASAKO N. DARROUGH AND
ROBERT H. BLANK

We hold these truths to be self-evident, that all men are
created equal, that they are endowed by their creator with
certain unalienable rights, that among these are Life, Liberty
and the Pursuit of Happiness.
                                    Declaration of Independence

That all men are equal is a proposition to which, at ordinary
times, no sane individual has even given his assent.
                                    Aldous Huxley: *Proper Studies*

Man is simply the most formidable of all the beasts of prey,
and, indeed, the only one that preys systematically on its
own species.
                                    William James, Remarks at the
                                    Peace Banquet, October 7, 1904

In this collection of ten essays, a series of issues related to
the topic of "biological differences and social equality" have
been explored from various points of view. The collection is
interdisciplinary involving scholars from a wide range of fields.
The major issue is what social equality means in the presence
of biological differences among human beings. The ideal of
"equality" is deeply entrenched in our society. Equality among
the equals is an easily acceptable principle. It requires little
effort to support. On the other hand, it is not always evident
who the equals are. Furthermore, what if we have the unequals?

How do we deal with the unequals? Is there such a notion as equality among the unequals? Equality among the similars? What do they mean?

The book is composed of two sections. The first section is devoted to a theoretical discussion, whereas the second section is devoted to social policy implications of biological differences. The theoretical section attempts to develop a framework for analysis. In so doing, we ask such questions as follows. What are biological differences? How are they relevant to society at large? How do we reconcile biological differences with the concept of equality? Included in this section are five essays written by scholars from as many disciplines: biology, law, anthropology, political science, and philosophy. These authors address themselves to this broad theoretical issue from various perspectives.

In the opening chapter, entitled "The Reality and Significance of Human Races: A Biological Perspective," Richard A. Goldsby, a biologist by training, presents a brief discussion of biological differences among humans and, in particular, differences according to "race." A race is defined as a "breeding population of individuals." Four races are studied as "demonstratably distinct" populations with different frequency distributions of various biological traits. A central theme emerging from this discussion is that there is much more intraspecie variations among human beings than among other animals. In addition, even if intergroup (interracial) differences are apparent (in terms of the average differences), intragroup variation may also be quite large. So it is possible that two individuals who come from the same group can be more different than two individuals from two different races. In addition, some of the differences in gene frequency among these races might represent traits which confer no (social and political) advantage or disadvantage on the population. Even if they do, much of our social behavior is shaped by both hereditary and environmental factors. Implied by this observation, for example, is that the low performance on IQ tests by blacks cannot be attributed unambiguously to their genetic constitution. A lesson to be learned then is that the biological basis of social behavior is very complicated to untangle and any attempt to do so requires extreme care.

Nancy R. Hauserman, in "Search for Equity on the Planet

Difference," explores the meaning of equity and equality for human beings who are "rich in biological differences." The article starts with a brief description of a hypothetical "planet of sameness," where all "hupersons" are "identical" in all respects. In such a world the concept of equality is meaningless and will never become an issue. After a search, she arrives at a concept of equality which states that "all persons are entitled to equal (identical) respect for their personage . . . irrespective of their abilities and achievements." In addition, she rejects any biological differences as the basis for greater shares of respect and hence for increased access to social rewards. There may be other reasons for inequality, but "there are no valid or relevant factors (which are biological) which should affect the consideration of each person as a human being." Thus, she concludes, equality is only a starting point, rather than an end in itself.

Thus we are thrown back to our starting point: how do we deal with biological differences in allocating scarce resources, if these biological differences affect the person's contribution (measured according to some criteria) to society? Perhaps it is time for us to seek wisdom (enlightenment) from scholars of the last few centuries. Three controversial but major developments in biological and social philosophy are discussed in "Evolution, Ethics, and Equality" by Steven L. Zegura, Stuart C. Gilman, and Robert L. Simon. Here an attempt is made by a physical anthropologist, a political scientist, and a philosopher to explain "how ideas from evolutionary biology, social Darwinism, and sociobiology might relate to ethical theory and the issue of equality."

The evolutionary theory promulgated by Darwin focused on the changes in the biological world brought about by changes in gene frequency due to mutation, natural selection gene flow, and genetic drift. On the other hand, social Darwinism, which is rooted in the works of Herbert Spencer, gave social and historical interpretation to Darwin's evolutionary theory. Sociobiology is a relatively new attempt to explain the social behavior of organisms in biological terms, based upon Darwinian evolutionary theory and strengthened by the hypothesis of "inclusive fitness" due to Hamilton. After a careful scrutiny of these ideas, the authors conclude that "even if complex human be-

haviors are genetically induced, they need not be impervious
to social change or moral criticism. . . . Rather than revealing that
selfish genes are our sovereign masters, a proper interpretation
of sociobiology suggests that it is we, ourselves, who can and
should determine the rightful boundaries of their domain."

Moving the discussion closer to the contemporary political
debate, Stuart C. Gilman examines "the development of a
specific view of equality," originated by Spencer and revived
by so-called new conservatives. His essay entitled "The Issue
of Inequality: Herbert Spencer and the Politics of the New
Conservatives" is a study in the politics of the sociology of
knowledge. The new conservatives (or old libertarians) argue
that "meritocracy is both the moral and the natural state of
man," that "man is, by his nature, intellectually predestined
to succeed or fail within the context of this meritocratic society,
and that modern American—or Western—society replicates this
status or at least comes close." That is, the contemporary American system of government, although it may not be the "best
of all possible worlds," is good enough to be defended. What
is at issue here is the trade-off between freedom and equality.

Gilman traces the above arguments back to the writings of
Spencer. Spencer, who defined the term "evolution" even
before Darwin, also viewed inequality to be the natural state
of man. It is the exceptional individual who is the driving force
behind progress, and to interfere with the process of natural
(social) evolution is to invite disaster. In modern social sciences,
Gilman claims most questions "can ultimately be understood
in terms of equality." When confronted with contradictions,
liberalism must turn either to the root-metaphor of Spencer or
the other intellectual giant of our times, Marx. In conclusion,
Gilman warns us that "a fundamental reevaluation of social
scientific work is necessary in order to view its ethical impact."

In the last chapter of the theoretical section titled "The
Liberal Conception of Equal Opportunity and Its Egalitarian
Critics," Robert L. Simon investigates the liberal or meritocratic position on "equal opportunity" which is probably the
most widely accepted interpretation of the concept of equality
in the contemporary American society. Equal opportunity is
defined as "a relationship holding among individuals when and

only when outcomes are determined in a nondiscriminatory fashion" (regardless of such factors as sex, race, or creed) and only by "fair competition." This, however, inevitably leads to an unequal outcome, and that is what troubles egalitarian critics such as John Schaar. They argue that the competitive outcomes themselves depend upon an arbitrary distribution of talents and abilities which are determined by factors beyond the individual's control. Ultimately one's talent and ability is controlled by genetic differences. The liberals, on the other hand, argue that meritocracy will result in more productive efficiency. And this concern about productive efficiency in turn will put greater emphasis on one's control over one's life and will broaden "the realm in which [one] can function as autonomous agent." Simon argues that the liberal meritocratic conception of equal opportunity "is not ascriptive, need not promote gross inequality of condition, and is not necessarily supportive of the status quo." Rather "it provides the rationale for condemnation of many unjust and outrageous inequalities that exclude all too many individuals from full participation in our society." On the other hand, the liberal's concept of equal opportunity does legitimize certain inequalities of outcome. To simply assume that the point of equal opportunity is to promote equal result, is only to confuse what is at stake. In fact, unequal results can be arrived at fairly under justifiable conditions. In summary, equal opportunity is "just the equal chance to influence outcomes through our choices, character, and capacities expressed in action." On the other hand, where competition is biased by the effects of past discrimination, deprivation, and injustice, such as in this society, Simon suggests that some form of corrective or affirmative action may well be compatible with the liberal's notion of equal opportunity.

Presented thus far were five essays which discussed human biological variation and offered several philosophical perspectives on the concept of social equality. The second section attempts to examine the interaction of biological differences and social equality in various policy areas. It focuses on specific policy concerns emerging from differences among humans within the context of the theoretical considerations relating to equality. Together these five essays demonstrate the breadth of social dimensions of biological differences in American society and the implications for social policy.

At the center of all social policy decisions are variations inherent in the population. Each policy tends to benefit some groups more than others and to deprive still others. While the social complexity of large, heterogeneous populations obscures biological differences at times, frequently these distinctions become the basis of inequalities. Certainly, much of the existing social, economic, and political discrimination falls along lines of biological differences. Sex, race, and ethnic background are among the most emphasized social variables, along with various socioeconomic and demographic variables. The biological variables, however, are permanent and in many cases more visible, possibly leading to institutionalized inequities. As a result, the biological differences can be tied closely to variation in many secondary socioeconomic variables such as income and education thus considerably confusing the situation.

In addition to the more emphasized differences of race and sex, there are differences based on genetic disorders. As prenatal diagnosis and genetic screening techniques advance, alternative policies will become available in our treatment of those affected with various types of genetic disorders. Genetic "abnormality" continues to be an obvious basis for discrimination and stigmatization. Advances in human genetic technology will not only influence our perceptions of the biological inequities but also have broad implications for social policies directed toward those with genetic disorders. At the forefront of these technologies are various screening and diagnostic techniques. Beyond is a wide array of more direct means of human genetic intervention. Basic to the acceptance or rejection of these techniques are society's conceptions of the desirability of biological diversity.

While it is commonly agreed among social observers that biological variations should not be the basis for invidious discrimination, these differences continue to be the most visible and persistent of factors leading to social and economic inequities. An underlying theme in the contributions to this section suggests that biological differences and their implications for social policy must be explicated in order to understand more fully the context of current social policy, and move toward amelioration of inequities based on biological distinctions among humans. Therefore, each of the essays in this section to some extent analyzes the social policy implications of particular biological distinctions as applied within the social and cultural context. The interplay between human biological differences and the philosophical frameworks of equality is obvious in these papers, despite the diversity of the policy implications they explore.

In her essay on "Biological Differences and Economic Equality: Race and Sex," Masako N. Darrough, an economist, attempts to integrate sociobiology, genetics, and economic theory in order to investigate economic inequality which exists among human beings; in particular, among different racial groups and between men and women. Although the concepts of race and sex are biologically originated, their implications on social and economic dimensions can be significantly different. Two important asymmetries which exist between race and sex are in terms of the pattern of capital accumulation and transfer and in terms of the reproductive roles between men and women. She concludes that policies designed to correct past inequities based on race or sex must recognize these asymmetries, and she briefly discusses the distinctive implications of each.

Sandra L. Schultz in "Social Equality and Ethnic Identity: A Case of Greek-Americans" extends the concept of biological difference to the concept of ethnic group. An anthropologist, Schultz emphasizes the cultural aspects of ethnic identity and discusses the role of component groups in a plural society. She is most disturbed by the policy orientation which assumes homogeneity within ethnic groups and thereby ignores intragroup differences. She reiterates the theme introduced by Goldsby that intragroup variation is normally higher than intergroup differences. Incorrectly, policy makers tend to focus on aggregate group differences while minimizing the heterogeneity of the members of the groups as a whole.

Through the Greek-American case, Schultz demonstrates substantial intragroup variation along generational, regional, and class lines. Policy makers must reevaluate their perceptions of ethnic group membership to account for this heterogeneity in order to assure social justice, according to Schultz, since these group subdivisions can and often do lead to inequality within the group.

Shifting the focus of attention to a question of bioethics, Nobel Laureate Baruch S. Blumberg, in "Social Equality and Infectious Disease Carriers," illustrates situations in which public health interests may be in conflict with individual liberty. Issues involving abrogation of personal rights may arise when individuals are identified as carriers of infectious diseases. The discussion centers around the case of the hepatitis B virus (or Australia antigen). Carriers of HBV are not necessarily sick themselves but are thought to be capable of contaminating others. Although not precisely known, the probability of transmission is likely to be rather small. But once identified, these carriers can be subject to stigmatization (i.e., discrimination, loss of employment, etc.) for fear of contamination. Unfortunately these carriers cannot be treated. The author points out that "there is a very wide discrepancy between the potential but unknown benefits to society of restricting carriers and the real hazards to the individuals," and concludes that ethical questions often cannot be resolved without further scientific knowledge, the type of knowledge which is the outgrowth of basic nondirected research.

In "Societal Responsibility and Genetic Disease: Some Political Considerations," Robert H. Blank shifts emphasis to biological variations defined by society as "disorders" or "diseases." He asks to what extent a democratic society has the responsibility to take steps necessary to reduce the occurrence of genetic disease and thereby reduce inequalities based on such genetic variations. He examines the political and social dilemmas raised by recent advances in human genetic technology. It is argued that each method of genetic intervention must be analyzed within the context of crucial social and political implications. Central to this discussion is a delineation of the conditions under which some social good might preempt certain individual rights and the extent to which genetic disease is a public health concern. Specific examples are drawn from genetic screening applications as well as several eugenic approaches.

In the last selection, John G. Borkowski, a psychologist, discusses an aspect of race differences which is central to the contemporary social policy debate. In his essay "Intelligence Theory and the Race-IQ Controversy," Borkowski reexamines perhaps the most sensitive of the issues surrounding biological differences and social equality. After reviewing traditional models of intelligence, he reanalyzes the race-IQ data using a revised, cognitive-based view of intelligence and finds that much of the point spread in IQ scores for black and white children is "attributable to the functioning of the executive system as it influences responses to complex problems." Borkowski reiterates the need to elevate information-processing skills for black children and suggests that "large racial differences in skills of the executive system exist primarily insofar as environmental stimulation . . . favors the members of one racial group over another."

The purpose of this concluding chapter is to draw together the major themes and ideas expressed in the substantive chapters and explicate some of the implications of biological differences in contemporary American society. Thus far we have attempted to provide a brief outline and summary of each of ten essays in this collection. The common theme has been equality in the face of biological differences, although different aspects of the theme and different approaches may have been taken in each essay. What then can we say in concluding this volume? What is clear is that the question of biological differences and social equality will be with us for a long time. The question is not easily solvable, partly due to lack of relevant information (see Blumberg), but mainly due to our lack of consensus as to what social equality means and how to achieve it. In a pluralistic society with vested interest groups, it is difficult to reach consensus on issues which involve transfers of power and wealth among the various groups. Of course it is this possible divergence in opinions which we as a society also value highly.

Unfortunately, the problems examined in these essays seem bound to be exacerbated in the coming decades, especially if particular social trends continue in the United States. Demographic shifts in population by age, reinforced by the probable displacement of large numbers of workers by technological developments and continuing pressures involving immigration of "aliens" (often easily identifiable by race), portend intensified rather than lessened emphasis on biological differences. Heightened competition among groups for constrained public services and a perceived trend toward intolerance of those who "are different" by some segments in society are certainly not encouraging signs in reducing the salience of biological difference

In addition, as biological and medical sciences make further progress, we expect that dilemmas relating to biological differences and their implications on social equality will become even more intricate. This is because with more sophisticated technology, some of the once-inevitable differences we have had are going to be subject to our manipulation. Clearly great potential benefits lie ahead of us. However, on the other hand, the problems of what, who, and how will not be resolved without serious deliberations. We will gain more degrees of apparent freedom but only at the cost of more difficult choices to make.

Together these essays urge caution in accepting oversimplified conceptions of biological variation among human beings. Biological models of the human species are more complex than those of other species. Intragroup differences must be recognized as at least as important for social policy as variation across biological or social groupings. Also, while we clearly exhibit genuine biological differences, we share as human beings more in common than we differ in our genetic constitution. If we are to advance as a community of humans, it is essential to recognize our common biological heritage and to work to minimize the impact of those biological differences which are deemed irrelevant on social policy decisions.

The issues raised by biological differences within the context of social equality are far more complicated than described by those individuals and groups vocal in the debate. More research is necessary to understand even the basic questions such as the ones addressed in this volume. This research, however, must be predicated upon a more coherent and comprehensive model of coexistence of various groups (and individuals) in society and of social equality among them. While there is a continuing need for interdisciplinary examinations in order to clarify the multipl dimensions of each of these concepts, considerable attention must be directed at the same time toward developing a dialogue which combines the theoretical discussion and its policy implications. We hope this volume proves to be successful at such an attempt.

Bibliography

Abernathy, G. L. (1959). *The Idea of Equality: An Anthology.* Richmond: John Knox Press.

Alter, H. J.; Chalmers, T. C.; Freeman, B. M.; Lunceford, J. L.; Lewis, T. L.; Holland, P. V.; Pizzo, P. A.; Plotz, P. H.; and Meyer, W. K. (1975). "Health-Care Workers Positive for Hepatitis B Surface Antigen. Are Their Contacts at Risk?" *New England Journal of Medicine* 292:454-57.

Andreski, Stanislav, ed. (1975). *Herbert Spencer: Structure, Function and Evolution.* London: Joseph.

Atkinson, Anthony B. (1973). *The Economics of Inequality.* Oxford: Clarendon Press.

Bachelder, B. L., and Denny, M. R. (1977). "A Theory of Intelligence: I. Span and the Complexity of Stimulus Control." *Intelligence* 1:127-50.

Barash, David P. (1977). *Sociobiology and Behavior.* New York: Elsevier.

Barzun, J. (1958). *Darwin, Marx, Wagner: Critique of a Heritage.* Garden City: Doubleday.

Baughman, E. E., and Dahlstrom, W. G. (1968). *Negro and White Children: A Psychological Study in the Rural South.* New York: Academic Press.

Becker, Gary S. (1976). "Altruism, Egoism, and Genetic Fitness: Economics and Sociobiology." *Journal of Economic Literature* 14(3):817-26.

———(1977). "Reply to Hirshleifer and Tullock." *Journal of Economic Literature* 15(2):506-7.

Bedau, Hugo Adam (1967). "Egalitarianism and the Idea of Equality." *Equality.* Edited by Pennock and Chapman. New York: Atherton Press.

Beecher, H. K. (1968). "Medical Research and the Individual." *Life or*

Death: Ethics and Options. Edited by D. H. Labby. Portland: Reed College.

Bell, Daniel (1972). "On Meritocracy and Equality." *The Public Interest* 29:29-68.

———(1974). "The Public Household-On 'Fiscal Sociology' and the Liberal Society." *The Public Interest* 37:29-68.

———(1975). "Ethnicity and Social Change." *Ethnicity: Theory and Experience.* Edited by Nathan Glazer and Daniel P. Moynihan. Cambridge: Harvard University Press.

Belmont, J. M. (1978). "Individual Differences in Memory: The Cases of Normal and Retarded Development." *Aspects of Memory.* Edited by M. Gruneberg and P. Morris. London: Methuen.

———, and Butterfield, E. C. (1977). "The Instructional Approach to Developmental Cognitive Research." *Perspectives on the Development of Memory and Cognition.* Edited by R. Kail and J. Hagen. Hillsdale, N. J.: Erlbaum Associates.

Benn, Stanley I. (1967). "Egalitarianism and Equal Consideration of Interests." *Equality.* Edited by Pennock and Chapman. New York: Atherton Press.

Berman, Marshall (1970). *The Politics of Authenticity.* New York: Atheneu

Bernstein, Richard (1976). *The Restructuring of Social and Political Theor* New York: Harcourt, Brace and Jovanovich.

Bessman, S. P., and Swazey, J. P. (1971). "PKU: A Study of Biomedical Legislation." *Human Aspects of Biomedical Innovation.* Edited by E. Mendelsohn. Cambridge: Harvard University Press.

Bird, Caroline (1971). *Born Female.* Revised ed. New York: Pocket Books.

Birren, J. E. (1974). "Translations in Gerontology—From Lab to Life: Psychophysiology and Speed of Response." *American Psychologist* 29:808-15.

Blumberg, B. S. (1976). "Bioethical Questions Related to Hepatitis B Antigen." *American Journal of Clinical Pathology* 65:848-53.

———(1977). "Australian Antigen and the Biology of Hepatitis B." *Science* 197:17-25.

———, and London, W. T. (1980). "Hepatitis B. Virus and Primary Hepatocellular Carcinoma: The Relation of 'Icrons' to Cancer." *Viruses in Naturally Occurring Cancers: Cold Spring Harbor Conferences on Cell Proliferation, VII.* Edited by M. Essex, G. Todaro, and H. Zurhausen. Cold Spring Harbor, New York: Cold Spring Harbor Laborator

———; London, W. T.; Sutnick, A. I.; Camp, F. R., Jr.; Luzzio, A. J.; and Conte, N. F. (1974). "Hepatitis Carriers among Soldiers who have returned from Vietnam: Australia Antigen Studies." *Transfusion* 14:63-66.

Bodmer, Walter F., and Cavalli-Sforza, Luigi Luca (1976). *Genetics, Evolution, and Man.* San Francisco: W. H. Freeman.

Borkowski, J. G. (1980). "Signs of Intelligence: Strategy Generalization and Metacognition." *The Development of Reflection.* Edited by S. Yussen. New York: Academic Press.

———, and Cavanaugh, J. C. (1979). "Maintenance and Generalization of Skills and Strategies by the Retarded." *Handbook of Mental Deficiency.* Edited by N. R. Ellis. 2nd ed. Hillsdale, N. J.: Erlbaum Associates.

Brace, C. L., and Montagu, M. F. A. (1965). *Man's Evolution.* New York: Macmillan.

Brand, C. (1980). "General Intelligence and Mental Speed: Their Relationship and Development." *Intelligence and Learning.* Edited by M. Friedman, J. P. Das, and N. O'Connor. London: Plenum Press.

Brewer, W. G. (1974). "There Is No Convincing Evidence for Operant or Classical Conditioning in Humans." *Cognition and the Symbolic Processes.* Edited by W. B. Weimer and D. S. Palermo. Hillsdale, N. J.: Erlbaum Associates.

Brittain, John A. (1977). *The Inheritance of Economic Status.* Washington, D. C.: The Brookings Institution.

Bronowski, J., and Mazlish, B. (1960). *The Western Intellectual Tradition from Leonardo to Hegel.* New York: Harper and Row.

Brown, A. L. (1978). "Knowing When, Where, and How to Remember: A Problem of Metacognition." *Advances in Instructional Psychology, 1.* Hillsdale, N. J.: Erlbaum Associates.

Buchanan, J. M. (1975). *The Limits of Liberty.* Chicago: The University of Chicago Press.

Buchler, J. (1951). *Toward a General Theory of Human Judgement.* New York: Columbia University Press.

Bunge, M. (1961). "The Weight of Simplicity in the Constructing and Assaying of Scientific Theories." *Philosophy of Science* 28:120-49.

Bunzel, John H. (1977). "Bakke vs. the University of California." *Commentary* 63(3):59-64.

Bureau of National Affairs (1973). *The Equal Employment Opportunity Act of 1972: Editorial Analysis, Discussion of Court Decisions Under 1964 Act, Text of Amended Act, Congressional Reports, Legislative History.* Washington, D. C.: Bureau of National Affairs.

Burt, C. (1949). "Subdivided Factors." *British Journal of Statistical Psychology* 2:41-63.

Butterfield, E. C., and Belmont, J. M. (1977). "Assessing and Improving the Cognitive Functions of Mentally Retarded People." *The Psychology of Mental Retardation: Issues and Approaches.* Edited by

I. Bialer and M. Sternlicht. New York: Psychological Dimensions.

Caldwell, M. B. (1954). "An Analysis of a Southern Urban Negro Population to Items on the Wechsler Intelligence Scale for Children." Unpublished doctoral dissertation, Pennsylvania State University.

——, and Smith, T. A. (1968). "Intellectual Structure of Southern Negro Children." *Psychological Reports* 23:63-71.

Callahan, D. (1973). *Tyranny of Survival: And Other Pathologies of Civilized Life.* New York: Macmillan.

Campbell, J. K. (1965). "Honour and the Devil." *Honour and Shame: The Values of Mediterranean Society.* Edited by J. G. Peristiany. London: Weidenfeld and Nicolson.

Campione, J. C., and Brown, A. L. (1978). "Toward a Theory of Intelligence: Contributions from Research with Retarded Children." *Intelligence* 2:279-304.

Caneiro, R. L. (1968). "Herbert Spencer." *International Encyclopedia of the Social Sciences.* Vol. XV. New York: Macmillan.

——(1973). "Structure, Function, and Equilibrium in the Evolutionism of Herbert Spencer." *Journal of Anthropological Research* 29(2):77-95.

Carlson, Lewis H., and Colburn, George A. (1972). *In Their Place: White America Defines Her Minorities, 1850-1950.* New York: Wiley.

Catlen, George E. G. (1967). "Equality and What We Mean by It." *Equality.* Edited by Pennock and Chapman. New York: Atherton Press.

Cattell, R. B. (1971). *Abilities: Their Structure, Growth, and Action.* Boston: Houghton Mifflin.

Cavalli-Sforza, Luigi Luca (1974). "The Genetics of Human Populations." *Scientific American* 231(September):80-89.

——, and Bodmer, Walter F. (1970). "Intelligence and Race." *Scientific American* 223(October):19-29.

——(1971). *The Genetics of Human Populations.* San Francisco: W. H. Freeman.

Chagnon, N. A., and Irons, William, eds. (1979). *Evolutionary Biology and Human Social Behavior: An Anthropological Perspective.* North Scituate, Mass.: Duxbury.

Chapman, John (1956). *Rousseau: Totalitarian or Democrat.* New York: Columbia University Press.

Christenson, Reo M. (1974). *Heresies Right and Left: Some Political Assumptions Reexamined.* New York: Harper and Row.

——(1976). *Challenge and Decision.* 5th ed. New York: Harper and Row.

Clark, A. S. (1923). "Correlation of the Auditory Digit Memory with General Intelligence." *The Psychological Clinic* 15:358-62.

Cohen, Marshall; Nagel, Thomas; and Scanlon, Thomas. (1977). *Equality and Preferential Treatment.* Princeton: Princeton University Press.

Coon, Carleton (1965). *The Living Races of Man.* New York: Alfred A. Knopf.

Coser, Lewis A., and Howe, Irving. (1977). *The New Conservatives.* New York: Signet.

Daniels, Norman (1978). "Merit and Meritocracy." *Philosophy and Public Affairs* 7(3):206-23.

Darwin, C. (1859). *On the Origin of Species.* London: John Murray.

——(1871). *The Descent of Man and Selection in Relation to Sex.* London: John Murray.

——(1872). *The Expression of the Emotions in Man and Animals.* London: John Murray.

Dawkins, R. (1976). *The Selfish Gene.* New York: Oxford University Press.

D'Entreves, A. P. (1967). *The Notion of State.* London: Oxford University Press.

Depres, L. A., ed. (1975). *Ethnicity and Resource Competition in Plural Societies.* The Hague: Mouton.

Dershowitz, A. M. (1976). "Karyotype, Predictability, and Culpability." *Genetics and the Law.* Edited by A. Milunsky. New York: Plenum Press.

Dobzhansky, Theodosius (1962). *Mankind Evolving.* New Haven: Yale University Press.

——; Ayala, F. J.; Stebbins, G. L.; and Valentine, J. W. (1977). *Evolution.* San Francisco: W. H. Freeman.

Dolbeare, Kenneth, and Dolbeare, Patricia. (1971). *American Ideologies.* Chicago: Markham.

Douglas, B. (1976). "The Common Good and the Public Interest." Paper presented at Northwestern University.

Du Boulay, Juliet (1974). *Portrait of a Greek Mountain Village.* Oxford Clarendon.

Dunn, L. C. (1965). *A Short History of Genetics.* New York: McGraw-Hill.

Ehrlich, P. R., and Feldman, S. S. (1977). *The Race Bomb: Skin Color, Prejudice, and Intelligence.* New York: Quadrangle.

Ellis, N. R., ed. (1970). "Memory Processes in Retardates and Normals." *International Review of Research in Mental Retardation.* Vol. IV. New York: Academic Press.

——(1979). *Handbook of Mental Deficiency.* 2nd ed. Hillsdale, N. J.: Erlbaum Associates.

Encyclopedia of the Social Sciences, Vol. V (1931). New York: Macmillan.

Epstein, A. L. (1978). *Ethos and Identity: Three Studies in Ethnicity.* The Hague: Mouton.

Fairchild, Henry P. (1911). *Greek Immigration to the United States.* New

Haven: Yale University Press.

Feinberg, J. (1974). *Doing and Deserving: Essays on the Theory of Responsibility.* Princeton: Princeton University Press.

Flathman, Richard E. (1967). "Equality and Generalization: A Formal Analysis." *Equality.* Edited by Pennock and Chapman. New York: Atherton Press.

Flavell, J. H. (1978). "Metacognitive Development." *Structural/Process Theories of Complex Human Behavior.* Edited by J. M. Scandura and C. J. Brainerd. Alphen a.d. Rijin, The Netherlands: Sijthoff and Noordhoff.

Fletcher, J. (1974). *The Ethics of Genetic Control: Ending Reproductive Roulette.* Garden City: Doubleday.

Flew, A. G. N. (1967). *Evolutionary Ethics.* New York: St. Martin's Press.

Foucault, Michel. (1972). *Archaeology of Knowledge.* New York: Harper and Row.

Frankel, Charles (1971). "Equality of Opportunity." *Ethics* 81(3):191-211

Friedl, Ernestine (1962). *Vasilika: A Village in Modern Greece.* New York: Holt, Rinehart and Winston.

Friedman, J. M. (1974). "Legal Implications of Amniocentesis." *University of Pennsylvania Law Review* 123:92-156.

Fuller, J. L. (1954). *Nature and Nurture: A Modern Synthesis.* New York: Doubleday.

Garber, H., and Herber, R. (1973). "The Milwaukee Project: Early Intervention as a Technique to Prevent Mental Retardation." *University of Connecticut Technical Papers.*

Garn, Stanley (1965). *"Human Races."* Springfield, Ill.: Thomas.

Geertz, Clifford, ed. (1963). *Old Societies and New States: The Quest for Modernity in Asia and Africa.* New York: Free Press.

——(1980). "Sociosexology." *New York Review of Books* 26(21-22):3-4.

Ghiselin, Michael T. (1978). "The Economics of the Body." *The American Economic Review* 68(2):233-37.

Gilewski, M. J.; Zelinski, E. M.; and Thompson, L. W. (1978). "Remembering Forgetting: Age Differences in Metamemorial Processes." Paper presented at the Annual Convention of the APA in Toronto, Canada, August, 1978.

Gilman, Charlotte Perkins (1966). *Women and Economics.* New York: Harper and Row.

Gilman, S. L. (1979). "Darwin Sees the Insane." *Journal of the History of the Behavioral Sciences* 15:253-62.

Gizelis, Gregory (1974). *Narrative Rhetorical Devices of Persuasion: Folklore Communication in a Greek-American Community.* Athens, Greece: National Centre of Social Research.

Glazer, Nathan (1975). *Affirmative Discrimination: Ethnic Inequality and Public Policy.* New York: Basic Books.

Golding, M. P. (1968). "Ethical Issues in Biological Engineering." *UCLA Law Review* 15:443-79.

Goldman, Alan H. (1977). "Justice and Hiring by Competence." *The American Philosophical Quarterly* 14(1):17-28.

Goldsby, Richard A. (1977). *Race and Races.* 2nd ed. New York: Macmillan.

Gordon, Scott (1979). "A Critique of Sociobiology." Discussion Paper no. 346, Queen's University, Kingston, Ontario, Canada.

Goudge, T. A. (1961). *Ascent of Life: A Philosophical Study of the Theory of Evolution.* Toronto: University of Toronto Press.

Gould, S. J. (1978). "Review of *On Human Nature* by E. O. Wilson," *Human Nature* 10:20-28.

Greeley, Andrew M. (1971). *Why Can't They Be Like Us? America's White Ethnic Groups.* New York: E. P. Dutton.

———(1974). *Ethnicity in the United States: A Preliminary Reconnaissance.* New York: John Wiley and Sons.

Greene, J. C. (1959). *The Death of Adam.* Ames: Iowa State University Press.

Grimmett, S. A. (1975). "Black and White Children's Free Recall of Unorganized and Organized Tests: Jensen's Level I and Level II." *Journal of Negro Education* 54:24-33.

Guilford, J. P. (1967). *The Nature of Human Intelligence.* New York: McGraw-Hill.

———(1979). "Intelligence Isn't What It Used to Be: What to Do About It." *Journal of Research and Development in Education* 12:33-46.

Gustafson, J. M. (1973). "Genetic Engineering and the Normative View of the Human." *Ethical Issues in Biology and Medicine.* Edited by P. N. Williams. Cambridge: Schenkman.

Guthrie, R. (1973). "Mass Screening for Genetic Disease." *Medical Genetics.* Edited by V. A. McKusick. New York: H. P. Publishing Company.

Haldane, J. B. S. (1932). *The Causes of Evolution.* New York: Longmans Green.

———(1955). "Population Genetics." *New Biology*, London, 18.

Hall, V. C., and Kaye, D. B. (1977). "Pattern of Early Cognitive Development Among Boys in Four Subcultural Groups." *Journal of Educational Psychology* 69:66-87.

Hamilton, W. D. (1964). "The Genetical Evolution of Social Behavior I and II." *Journal of Theoretical Biology* 7:1-52.

Harris, Marvin (1968). *The Rise of Anthropological Theory.* New York: T. Y. Crowell.

———(1980). "Sociobiology and Biological Reductionism." *Sociobiology*

Examined. Edited by Ashley Montagu. Oxford: Oxford University Press.

Harris, R. J. (1960). *The Quest for Equality.* Baton Rouge: Louisiana State University Press.

Hartley, L. P. (1960). *Facial Justice.* London: Morrison and Gibb.

Heidegger, Martin. (1961). *An Introduction to Metaphysics.* New York: Anchor.

Heim, W. G. (1973). "Moral and Legal Decisions in Reproductive and Genetic Engineering." *Ethical Issues in Biology and Medicine.* Edited by P. N. Williams. Cambridge: Schenkman.

Heisenberg, W. (1962). *Physics and Philosophy.* New York: Harper Torch Books.

Herrnstein, Richard (1971a), "IQ." *The Atlantic Monthly* 228(3):43-64.

———(1971b). "Professor Herrnstein Replies." *The Atlantic Monthly* 228(6):110.

Hess, R. D., and Shipman, V. C. (1965). "Early Experience and the Socialization of Cognitive Modes in Children." *Child Development.* 36:869-86.

Hiestand, Dale L. (1970). "Discrimination in Employment: An Appraisal of the Research Policy." *Papers in Human Resources in Industrial Relations 16.* The Institute of Labor and Industrial Relations. Ann Arbor: University of Michigan.

Hirsch, Jerry (1975). "Jensenism: The Bankruptcy of Science." *Educational Theory* 25(1):3-27.

———(1976). "An Epitaph for Sir Cyril?" *Newsweek,* Dec. 20:76.

———, et al. (1975). "Race, Class, and Intelligence: A Critical Look at the IQ Controversy." *International Journal of Mental Health* 3(4):46-142.

Hirshleifer, J. (1977a). "Economics from a Biological Viewpoint." *Journal of Law and Economics* 20(1):1-52.

———(1977b). "Shakespeare vs. Becker on Altruism: The Importance of Having the Last Word." *Journal of Economic Literature* 15(2):500-50?

———(1978). "Competition, Cooperation, and Conflict in Economics and Biology." *American Economic Review* 68(2):238-43.

Holloman, Regina E., ed. (1978). *Perspectives on Ethnicity.* The Hague: Mouton.

Hook, S., ed. (1964). *Law and Philosophy: A Symposium.* New York: New York University Press.

Horn, J. L. (1968). "Organization of Abilities and the Development of Intelligence." *Psychological Review* 75:242-59.

———(1978). "The Nature and Development of Intellectual Abilities." *Human Variation.* Edited by R. T. Osborne, C. E. Noble, and N. Weyls. New York: Academic Press.

Horowitz, Donald L. (1975). "Ethnic Identity." *Ethnicity: Theory and*

Experience. Edited by Nathan Glazer and Daniel P. Moynihan. Cambridge: Harvard University Press.

Howard, John R., ed. (1972). *Awakening Minorities: American Indians, Mexican Americans, Puerto Ricans.* Chicago: Aldine.

Hubbard, R. (1978). "From Termite to Human Behavior; Reflections on Sociobiologist Edward Wilson's *On Human Nature." Psychology Today* 12(5):124-35.

Hull, D. (1969). "What Philosophy of Biology Is Not." *Journal of the History of Biology* 2:241-68.

Jencks, Christopher, et al. (1972). *Inequality.* New York: Basic Books.

Jensen, A. R. (1969). "How Much Can We Boost IQ and Scholastic Achievement?" *Harvard Educational Review* 39:1-123.

———(1979). "g: Outmoded Theory or Unconquered Frontier." *Creative Science and Technology* 2:16-29.

———, and Frederiksen, J. (1973). "Free Recall of Categorized and Uncategorized Tests: A Test of the Jensen Hypothesis." *Journal of Educational Psychology* 65:304-12.

———, and Figueroa, R. A. (1975). "Forward and Backward Digit Span Interaction with Race and IQ: Predictions from Jensen's Theory." *Journal of Educational Psychology* 67:882-93.

Johnson, Harry Alleyn (1976). *Ethnic American Minorities: A Guide to Media and Materials.* New York: R. R. Bowker.

Kaback, M. M., and O'Brien, J. S. (1973). "Tay-Sachs: Prototype for Prevention of Genetic Disease." *Medical Genetics.* Edited by V. A. McKusick. New York: H. P. Publishing Company.

Kamin, Leon (1974). *The Science and Politics of IQ.* Maryland: Erlbaum Associates, distributed by John Wiley and Sons, New York.

———(1976). "Heredity, Intelligence, Politics, and Psychology: II." *The IQ Controversy.* Edited by N. J. Block and Gerald Dworkin. New York: Pantheon Books.

Karp, L. E. (1976). *Genetic Engineering: Threat or Promise.* Chicago: Nelson Hall.

Kavka, Gregory (1976). "Equality in Education." *Indeterminacy in Education.* Edited by John E. McDermott. Berkeley: McCutchan.

Keller, A. G., ed. (1914). *The Challenge of Facts and Other Essays,* New Haven: Yale University Press.

Kendall, C., Borkowski, J. G., and Cavanaugh, J. C. (1980). "Maintenance Generalization of an Interrogative Strategy by EMR Children." *Intelligence* 4:255-70.

Kennedy, Wallace A.; Van de Rait, Vernon; and White, James C., Jr. (1963). "A Normative Sample of Intelligence and Achievement of Negro Elementary School Children in the Southeastern United States." *Monographs of the Society for Research in Child Development* 28(6).

Kieffer, G. (1974). *Ethical Issues in the Life Sciences.* New York: American Council of Learned Societies.

King, J. L., and Jukes, T. H. (1969). "Non-Darwinian Evolution." *Science* 164:788-98.

Kourvetaris, George A. (1971). *First- and Second-Generation Greeks in Chicago.* Athens, Greece: National Centre of Social Research.

———(1976). "The Greek-American Family." *Ethnic Families in America: Patterns and Variations.* Edited by Charles H. Mindel and Robert W. Habenstein. New York: Elsevier.

Kreutzer, M. A.; Leonard, C.; and Flavell, J. H. (1975). "An Interview Study of Children's Knowledge about Memory." *Monographs of the Society for Research in Child Development* 40(serial no. 159).

Kuhn, T. (1970). *The Structure of Scientific Revolutions.* Chicago: University of Chicago Press.

Kuper, Leo, and Smith, M. G., eds. (1969). *Pluralism in Africa.* Berkeley: University of California Press.

Lakoff, Stanford A. (1964). *Equality in Political Philosophy.* Boston: Beacon Press.

Lappé, M. (1972). "Moral Obligations and the Fallacies of Genetic Control." *Theological Studies* 33:411-27.

———(1975). "Can Eugenic Policy Be Just?" *The Prevention of Genetic Disease and Mental Retardation.* Edited by A. Milunsky. Philadelphia: W. B. Saunders Company.

———, Gustafson, J. M.; and Roblin, R. (1972). "Ethical and Social Issues in Screening for Genetic Disease." *New England Journal of Medicine* 286:1129-32.

Laslett, P., ed. (1967). *Philosophy, Politics, and Society.* Oxford: Basil Blackwell.

Leber, George J. (1972). *The History of the Order of Ahepa (The American Hellenic Educational Progressive Association): 1922-1972. Including the Greeks in the New World, and Immigration to the United States.* Washington, D. C.: The Order of the Ahepa.

Lee, Dorothy (1959). *Freedom and Culture.* Englewood Cliffs: Prentice-Hall.

Lerner, I. M., and Libby, William J. (1976). *Heredity, Evolution, and Society.* 2nd ed. San Francisco: W. H. Freeman.

Lesser, G. S.; Fifer, G.; and Clark, D. H. (1965). "Mental Abilities of Children from Different Social Class and Cultural Groups." *Monographs of the Society for Research in Child Development* 30(4, serial no. 102):1-115.

Levenstein, P. (1970). "Cognitive Growth in Preschoolers Through Verbal Interaction with Mothers." *American Journal of Orthopsychiatry* 40:426-32.

———(1975). "Message from Home: Findings from a Program for Non-Retarded, Low-Income Preschoolers." *The Mentally Retarded and Society: A Social Science Perspective.* Edited by M. J. Begab and S. A. Richardson. Baltimore: University Park Press.

Levin, J. (1977). *Learner Differences: Diagnosis and Prescription.* New York: Holt Rinehart.

Levine, Irving M. (1971). "Government's Role in Meeting the Needs of White Ethnic Citizens." *Ethnic Groups in the City: Culture, Institutions, and Power.* Edited by Otto Feinstein. Lexington: Heath Lexington Books.

Levy, H. L. (1973). "Genetic Screening." *Advances in Human Genetics.* Vol. XIV. Edited by H. Harris and K. Hirshborn. New York: Plenum Press.

Lillard, Lee A. (1977). "Inequality: Earnings vs. Human Wealth." *American Economic Review* 67(2):42-53.

Lipset, Seymour Martin (1960). *Political Man.* New York: Anchor.

Litt, Edgar (1970). *Ethnic Politics in America: Beyond Pluralism.* Glenview: Scott, Foresman.

Lockhard, Joan S. (1980). *The Evolution of Human Social Behavior.* New York: Elsevier North Holland.

Loehlin, J. C.; Lindzey, G.; and Spuhler, J. N. (1975). *Race Differences in Intelligence.* San Francisco: W. H. Freeman.

London, W. T.; Drew, J. S.; Lustbader, E. D.; Werner, B. G.; and Blumberg, B. S. (1977). "Host Response to Hepatitis B Infection in Patients in a Chronic Hemodialysis Unit." *Kidney International* 12:51-58.

Lucas, J. (1978). *Ready, Set, Remember.* Nashville: Impact Press.

McCloskey, R. G. (1964). *American Conservatism in the Age of Enterprise.* New York: Harper and Row.

McCune, Shirley, and Matthews, Martha (1975). *Programs for Education Equality: Schools and Affirmative Action.* Washington, D. C.: U. S. Dept. of Health, Education, and Welfare, Office of Education.

McGlone, Jeanette (1980). "Sex Differences in Human Brain Asymmetry: A Critical Survey." *The Behavioral and Brain Sciences* 3:215-27.

Makielski, Stanislaw J., Jr. (1973). *Beleaguered Minorities: Cultural Politics in America.* San Francisco: W. H. Freeman.

Matras, J. (1975). *Social Inequality, Stratification, and Mobility.* New Jersey: Prentice-Hall.

Matsunaga, E. (1962). "The Dimorphism in Human Normal Cerumen." *Annals of Human Genetics* 25(4):273-86.

May, R. M. (1978). "The Evolution of Ecological Systems." *Scientific American* 239:160-75.

Mayr, E. (1978). "Evolution." *Scientific American* 239:46-65.

Meichenbaum, D., and Asarnow, J. (1978). "Cognitive-Behavior Modification and Metacognitive Development: Implications for the Class-room." *Cognitive-Behavioral Interventions: Theory, Research, and Procedures.* Edited by P. Kendall and S. Holton. New York: Academic Press.

Michelman, F. I. (1969-70). "On Protecting the Poor Through the Four-teenth Amendment." Foreword to "The Supreme Court: 1968 Term." *Harvard Law Review* 83:7-59.

Miller, D. (1976). *Social Justice.* London: Clarendon Press.

Milunsky, A., ed. (1976). "Medico-Legal Issues in Prenatal Genetic Diag-nosis." *Genetics and the Law.* New York: Plenum Press.

———, and Atkins, L. (1975). "Prenatal Diagnosis of Genetic Disorders." *The Prevention of Genetic Disease and Mental Retardation.* Edited by A. Milunsky. Philadelphia: W. B. Saunders Company.

Mindel, Charles, and Habenstein, Robert W., eds. (1976). *Ethnic Families in America: Patterns and Variations.* New York: Elsevier.

Montagu, Ashley (1963). *Race, Science, and Humanity.* New York: Van Nostrand Reinhold.

———(1964). *The Concept of Race.* Toronto: Collier Macmillan.

———, ed. (1980). *Sociobiology Examined.* Oxford: Oxford University Press.

Mosley, J. W. (1975). "The HBV Barrier—A New Kind of Leper?" *New England Journal of Medicine* 292:477-78.

Muller, H. J. (1959). "The Guidance of Human Evolution." *Perspectives in Biology and Medicine* 1:16-35.

Murray, R. F. (1972). "Problems Behind the Promise: Ethical Issues in Mass Genetic Screening." *Hastings Center Report* 2.

Myrdal, Gunnar (1944). *The American Dilemma.* New York: Harper and Row.

Nadler, H. L. (1972). "Prenatal Detection of Genetic Disorders." *Advances in Human Genetics.* Vol. III. Edited by H. Harris and K. Hirshhorn. New York: Plenum Press.

National Academy of Sciences (1975). *Genetic Screening: Programs, Principles, and Research.* Washington: NAS.

Neimark, E. D. (1976). "The Natural History of Spontaneous Mnemonic Activity Under Conditions of Minimal Experimental Constraint." *Minnesota Symposia on Child Psychology.* Vol. X. Edited by A. D. Pick. Minneapolis: University of Minnesota Press.

Newman, William M. (1973). *American Pluralism: A Study of Minority Groups and Social Theory.* New York: Harper and Row.

Nisbet, Robert (1973). *The Social Philosophers: Community and Conflict in Western Thought.* New York: Crowell.

———(1974). "The Pursuit of Equality." *The Public Interest* 35:103-20.

Novak, Michael (1972). *The Rise of the Unmeltable Ethnics: Politics and Culture in the Seventies.* New York: Macmillan.

Okun, Arthur M. (1975). *Equality and Efficiency: The Big Trade-off.* Washington, D. C.: The Brookings Institution.

O'Neil, H. F. (1978). *Learning Strategies.* New York: Academic Press.

Ornstein, P. A. (1978). *Memory Development in Children.* Hillsdale, N. J.: Erlbaum Associates.

Papanikolas, Z. Helen (1970). *Toil and Rage in a New Land: The Greek Immigrant in Utah.* Salt Lake City: Utah Historical Society.

Patterson, James (1970). "The Unassimilated Greeks of Denver." *Anthropological Quarterly* 43:243-53.

Peel, J. D. Y., ed. (1972). *Herbert Spencer on Social Evolution.* Chicago: University of Chicago Press.

Pellegrino, J. W., and Glaser, R. (1979). "Cognitive Correlates and Components in the Analysis of Individual Differences." *Intelligence* 3:187-214.

Pennock, J. R., and Chapman, J. W. (1967). *Equality.* New York: Atherton Press.

Pepper, S. (1966). *World Hypotheses.* Berkeley: University of California Press.

Perrin, R. G. (1976). "Herbert Spencer's Four Theories of Social Evolution." *American Journal of Sociology* 81(6):1339-59.

Peter, W. G. (1975). "Ethical Perspectives in the Use of Genetic Knowledge." *Human Genetics.* Edited by T. R. Mertens. New York: John Wiley and Sons.

Plamenatz, John (1967). "Diversity of Rights and Kinds of Equality." *Equality.* Edited by Pennock and Chapman. New York: Atherton Press.

Polanyi, Karl (1975). *The Great Transformation.* New York: Octagon.

Pole, J. R. (1978). *The Pursuit of Equality in American History.* Berkeley: University of California Press.

Politis, M. J. (1945). "Greek-Americans." *One America: The History, Contributions, and Present Problems of Our Racial and National Minorities.* Edited by Francis J. Brown and Joseph S. Roucek. New York: Prentice-Hall.

Popper, Karl (1950). *The Open Society and Its Enemies.* Revised ed. Princeton: Princeton University Press.

Posner, M. I., and Boies, S. J. (1971). "Components of Attention." *Psychological Review* 78:391-408.

Provine, W. B. (1973). "Geneticists and the Biology of Race Crossing." *Science* 182:790-96.

Ramsey, P. (1975). "Genetic Engineering." *Human Genetics.* Edited by T. R. Mertens. New York: John Wiley and Sons.

Rawls, John (1971). *A Theory of Justice*. Cambridge, Mass.: Harvard University Press.

Rees, John (1971). *Equality*. New York: Praeger.

Reichhart, G. J., and Borkowski, J. G. (1978). "Effects of Clustering Instructions and Category Size on Free Recall of Normal and Retarded Children." *American Journal of Mental Deficiency* 83:73-76.

Reilly, P. (1975). "Genetic Screening Legislation." *Advances in Human Genetics*. Vol. V. Edited by H. Harris and K. Hirshhorn. New York: Plenum Press.

———(1977). *Genetics, Law, and Social Policy*. Cambridge: Harvard University Press.

Rein, Martin (1970). *Social Policy: Issues of Choice and Chance*. New York: Random House.

———(1976). *Social Science and Public Policy*. New York: Penguin.

Resnick, L. B. (1976). *The Nature of Intelligence*. Hillsdale, N. J.: Erlbaum Associates.

Ricoeur, Paul (1969). *The Symbolism of Evil*. Boston: Beacon.

Rohwer, W. D. (1971). "Learning, Race, and School Success." *Review of Educational Research* 41:191-210.

Rosaldo, Michelle Zimbalist, and Lamphere, Louise, ed. (1974). *Women, Culture and Society*. Stanford: Stanford University Press.

Rose, Arnold M., and Rose, Caroline B. (1965). *Minority Problems*. New York: Harper and Row.

Rousseau, J. J. (1964). *The First and Second Discourses*. Edited by Roger D. Masters. Translated by Roger D. and Judith R. Masters. New York: St. Martin's Press.

Rubin, Vera, ed. (1960). "Social and Cultural Pluralism in the Caribbean." *Annals of the New York Academy of Sciences* 83.

Ryan, William (1976). *Blaming the Victim*. (Revised, updated ed.). New York: Vintage Books.

Safilios-Rothschild, Constantina (1967). "Class Position and Success Stereotypes in Greek and American Culture." *Social Forces* 45:374-82.

Sahlins, Marshall (1976). *The Use and Abuse of Biology*. Ann Arbor, Mich.: University of Michigan Press.

Saloutos, Theodore (1964). *The Greeks in the United States*. Cambridge: Harvard University Press.

Salthouse, T. A. (1979). "Age and Memory: Strategies for Localizing the Loss." *New Directions in Aging and Memory*. Edited by L. W. Poon, J. L. Fozard, L. S. Gernak, D. Arenberg and L. W. Thompson. Hillsdale, N. J.: Erlbaum Associates.

Salzinger, S.; Salzinger, K.; and Hobson, S. (1967). "The Effect of Syntactical Structure on Immediate Memory for Word Sequences in

Middle- and Lower-Class Children." *The Journal of Psychology* 67:147-59.

Samuelson, Paul A. (1978). "Maximizing and Biology." *Economic Inquiry* 16(2):171-83.

Sanders, Irwin T. (1962). *Rainbow in the Rock: The People of Rural Greece.* Cambridge: Harvard University Press.

Scarr, S., and Weinberg, R. A. (1976). "IQ Test Performance of Black Children Adopted by White Families." *American Psychologist* 31: 726-39.

Schaar, John H. (1967). "Equality of Opportunity and Beyond." *Equality.* Edited by Pennock and Chapman. New York: Atherton Press.

Schelling, T. C. (1978). *Micromotives and Macrobehavior.* New York: W. W. Norton.

Schratz, Marjorie M. (1978). "A Developmental Investigation of Sex Differences in Spatial (Visual-Analytic) and Mathematical Skills in Three Ethnic Groups." *Developmental Psychology* 14(3):263-67.

Schultz, Theodore W. (1975). "The Value of the Ability to Deal with Disequilibria." *Journal of Economic Literature* 13(3):827-46.

Schumach, Murray (1978). "Astoria, the Largest Greek City Outside Greece." *Ethnic America.* Edited by Marjorie Weiser. New York: Wilson.

Scriven, M. (1959). "Explanation of Prediction in Evolutionary Theory." *Science* 120:477-82.

Scrofani, P. J. (1972). "Effects of Pretraining the Conceptual Functions of Children at Different Levels of SES, IQ, and Cognitive Development." *Proceedings of the 80th Annual Convention of the APA* 7:427-28.

Sen, Amartya (1973). *On Economic Inequality.* Oxford: Clarendon Press.

Senior, J. R.; Sutnick, A. I.; Goeser, E.; London, W. T.; Dahlke, M. D.; and Blumberg, B. S. (1974). "Reduction of Posttransfusion Hepatitis by Exclusion of Australia Antigen from Donor Blood in an Urban Public Hospital." *American Journal of the Medical Sciences* 267:171-77.

Shibutani, Tamotsu, and Kwan, Kian M. (1965). *Ethnic Stratification: A Comparative Approach.* New York: Macmillan.

Shuey, A. M. (1966). *The Testing of Negro Intelligence.* New York: Social Science Press.

Simon, Robert L. (1973). "Equal Opportunity and the Serrano Decision." *Intellect* 102 (2351): 32-35.

———(1978-79). "An Indirect Defense of the Merit Principle." *The Philosophical Forum* 10(2-4):224-41.

Simpson, G. G. (1961). *Principles of Animal Taxonomy.* New York: Columbia University Press.

Sitkei, E. G., and Myers, C. E. (1969). "Comparative Structure of Intellect in Middle- and Lower-Class Four-Year-Olds of Two Ethnic Groups." *Developmental Psychology* 1:592-604.

Sokol, R. P. (1967). *Puzzle of Equality.* Charlottesville: The Michie Co.

Sowell, Thomas (1976). *Race and Economics.* New York: McKay.

Spencer, Herbert (1880). *The Data of Ethics.* New York: T. Y. Crowell.

———(1884). *Illustrations of Universal Progress.* New York: D. Appleton.

———(1896a). *The Principles of Ethics.* Vols. I and II. New York: D. Appleton.

———(1896b). *Social Statics.* New York: D. Appleton.

———(1896c). *The Study of Sociology.* New York: D. Appleton.

———(1896d). *Principles of Sociology.* Vols. I and II. New York: D. Appleton.

Stein, Howard F., and Hill, Robert F. (1977). *The Ethnic Imperative: Examining the New White Ethnic Movement.* University Park: Pennsylvania State University.

Stephanides, Marios (1971). "Detroit's Greek Community." *Ethnic Groups in the City: Culture, Institutions, and Power.* Edited by Otto Feinstein, Lexington: Heath Lexington Books.

Stephen, J. F. (1967). *Liberty, Equality, Fraternity.* New York: Henry Holt.

Stern, Curt (1973). *Principles of Human Genetics.* 3rd ed. San Francisco: W. H. Freeman.

Sternberg, R. J. (1979). "The Nature of Mental Abilities." *American Psychologist* 34:214-30.

———, and Detterman, D. K. (1979). *Human Intelligence.* Norwood, N. J.: Ablex.

Sternberg, S. (1966). "High Speed Scanning in Human Memory." *Science* 153:652-54.

Stevenson, H. W.; Parker, T.; Wilkinson, A.; Bonnevaux, B.; and Gonzalez, M. (1978). "Schooling, Environment, and Cognitive Development: A Cross-Cultural Study." *Monographs of the Society for Research in Child Development* 43(3, serial no. 175):1-78.

Stodolsky, S. S., and Lesser, G. (1967). "Learning Patterns in the Disadvantaged." *Harvard Educational Review* 37:546-93.

Stycos, Mayone J. (1948). "The Spartan Greeks of Bridgetown: The Second Generation." *Common Ground* 8:72-86.

Sumner, William G. (1963). "Socialism." Reprinted in Stow Persons, *Social Darwinism: Selected Essays of William Graham Sumner.* Englewood Cliffs, N. J.: Prentice-Hall.

Symons, Donald (1980). "Précis of the Evolution of Human Sexuality." *The Behavioral and Brain Sciences* 3:171-81.

Taubman, Paul (1975). *Sources of Inequality in Earnings.* Amsterdam: North-Holland.

Tavuchis, Nicholas (1972). *Family and Mobility Among Second-Generation Greek-Americans.* Athens, Greece: National Centre of Social Research.

Tawney, R. H. (1929). *Equality.* New York: Barnes and Noble.

Terman, Lewis (1916). *The Measurement of Intelligence.* Boston: Houghton Mifflin.

——(1917). "Feeble-minded Children in the Public Schools of California." *School and Society* 5:161-65.

——(1925). *Genetic Studies of Genius.* (4 volumes). Stanford: Stanford University Press.

Theibar, G. W., and Feldman, S. D. (1972). *Issues on Social Inequality.* Boston: Little Brown.

Theodoratus, Robert James (1971). "A Greek Community in America: Tacoma, Washington, Sacramento." *Anthropological Society Paper 10.* Sacramento.

Thomson, D. (1949). *Equality.* Cambridge: Cambridge University Press.

Thurstone, T. L. (1947). *Multiple-Factor Analysis.* Chicago: University of Chicago Press.

Tobin, J. (1958). "Liquidity Preference as Behavior Towards Risk." *The Review of Economic Studies* 25:65-86.

Trivers, R. L. (1971). "The Evolution of Reciprocal Altruism." *Quarterly Review of Biology* 46:35-57.

Tullock, Gordon (1977). "Economics and Sociobiology: A Comment." *Journal of Economic Literature* 15(2):502-6.

U. S. Commission on Civil Rights (1973). *Statement of Affirmative Action for Equal Employment Opportunities.* Publication 41. Washington, D. C.

Uzgiris, J. C., and Hunt, J. McV. (1974). *Toward Ordinal Scales of Psychological Development of Infancy.* Urbana, Ill.: University of Illinois Press.

Van den Berghe, Pierre L. (1980). "The Human Family." *The Evolution of Human Social Behavior.* Edited by Lockhard. New York: Elsevier North Holland.

Vernon, P. E. (1979). *Intelligence: Heredity and Environment.* San Francisco: W. H. Freeman.

Vlachos, Evangelos C. (1968). *The Assimilation of Greeks in the United States, with Special Reference to the Greek Community of Anderson, Indiana.* Athens, Greece: National Centre of Social Research.

Von Neuman, J., and Morgenstern, O. (1953). *The Theory of Games and Economic Behavior.* 3rd ed. Princeton: Princeton University Press.

Warner, W. Lloyd, and Srole, Leo (1945). *The Social System of American*

Ethnic Groups. New Haven: Yale University Press.

Washburn, S. L. (1978). "Human Behavior and the Behavior of Other Animals." *American Psychologist* 33:405-18.

Webster's New Collegiate Dictionary, 1949 ed.

Weed, Perry L. (1973). *The White Ethnic Movement and Ethnic Politics.* New York: Praeger.

Weinberg, M. (1977). *A Chance to Learn: The History of Race and Education in the U. S.* London: Cambridge University Press.

Weiss, Y. (1972). "The Risk Element in Occupational and Educational Choices." *Journal of Political Economy* 80(6):1203-13.

Welch, Finis (1976). "Ability Tests and Measures of Differences Between Black and White Americans." *Working Paper R2102.* Santa Monica, CA: Rand Corporation.

Wild, John (1953). *Plato and His Modern Enemies.* Chicago: University of Chicago Press.

Williams, Bernard (1971). "The Ideal of Equality." *Philosophy, Politics, and Society.* Edited by P. Laslett and W. G. Runciman (1962). London: Basil Blackwell. Reprinted in Hugo A. Bedau, ed. (1971). *Justice and Equality.* Englewood Cliffs, N. J.: Prentice-Hall.

Williams, Robin M., Jr. (1977). *Mutual Accommodation: Ethnic Conflict and Cooperation.* Minneapolis: University of Minnesota Press.

Wilson, Edward O. (1975). *Sociobiology, the New Synthesis.* Cambridge, Mass.: Harvard University Press.

———(1978). *On Human Nature.* Cambridge, Mass.: Harvard University Press.

Wilson, John (1966). *Equality.* New York: Harcourt, Brace and World.

Wysocki, B. A., and Wysocki, A. G. (1969). "Cultural Differences as Reflected in Wechsler-Bellevue Intelligence (WBII) Test." *Psychological Reports* 25:95-101.

Xenides, J. P. (1922). *The Greeks in America.* New York: George H. Doran.

Young, F. M., and Bright, H. A. (1954). "Results of Testing 81 Negro Rural Juveniles with the Wechsler Intelligence Scale for Children." *Journal of Social Psychology* 39:219-26.

———, and Pitts, V. A. (1951). "The Performance of Congenital Syphilitics on the Wechsler Intelligence Scale for Children." *Journal of Consulting Psychology* 15:239-42.

Name Index

Subject Index

About the Editors and Contributors

Robert H. Blank is Professor of Political Science at the University of Idaho. His research interests include political values and political implications of genetic research. Among his published works is *Political Implications of Human Genetic Technology* (Westview Press, 1981).

Baruch S. Blumberg is Associate Director for Clinical Research at the Institute for Cancer Research, University Professor of Medicine and Anthropology at the University of Pennsylvania, and a Nobel Laureate in Physiology of Medicine (1976). Although his major research interest lies in the biology of viruses and cancer, he also teaches medical anthropology using the experience from extensive field work in developing countries. He is the author of over three hundred articles.

John G. Borkowski is Professor of Psychology at the University of Notre Dame. He has been primarily interested in the psychology of cognitive development from life-span perspective. He is the author of *Experimental Psychology: Tactics of Behavioral Research* (with D. C. Anderson; Scott, Foresman, and Company, 1977) as well as numerous articles in the psychology area.

Masako N. Darrough is Assistant Professor in the Faculty of Commerce at the University of British Columbia. She has done research in widely ranging areas related to economics and business. Her published works have appeared in journals including *International Economic Review* and *Journal of Business.*

Stuart C. Gilman is Assistant Professor of Political Science and Professor of American Studies at Saint Louis University. He is currently involved in research on the understanding of equality in contemporary society and on the impact of the development of medicine on political thought. Among the books he has coauthored is *Race and Ethnic Relations* (with Reid Luhman; Wadsworth, 1980).

Richard A. Goldsby is Cross Professor of Biology at Amherst College. The focus of his research has been the somatic cell genetics of the immune system. He has written, among others, *Race and Races* (Macmillan, 1977) as an exposition of the biology of races.

Nancy R. Hauserman is Assistant Professor of Industrial Relations at the University of Iowa and an attorney. Her major research interest is on legal and social problems of homemakers. She has written articles on different topics in legal and business journals.

Gardner Lindzey is Director of the Center for Advanced Study in the Behavioral Sciences in California. He has taught psychology in various schools including Harvard University, University of Texas, and University of Minnesota. Among his major works are *Theories of Personalities* (with Calvin S. Hall; John Wiley and Sons, 1978) and *Race Differences in Intelligence* (with J. C. Loehlin and J. N. Spuhler; W. H. Freeman, 1975).

Kurt L. Schlesinger is Professor of Psychology and Fellow at the Institute for Behavioral Genetics, University of Colorado, Boulder. He has done research and published in the area of genetic and biochemical determinants of behavior.

Sandra L. Schultz is an anthropologist of broad interests. She has taught at various universities, has done applied work for the federal government, and is currently moving into the public health area. The focus of her interests, however, has been the acculturation and adaptation process of immigrant groups.

Robert L. Simon is William R. Kenan, Jr., Professor of Philosophy at Hamilton College. His interests are political and social philosophy as well as the philosophy of law. He is associate editor of *Ethics* and the coauthor of *The Individual and the Political Order* (with N. E. Bowie, Prentice-Hall, 1977).

Stephen L. Zegura is Associate Professor of Anthropology and Genetics at the University of Arizona. His interests center around physical anthropology as well as human biology and genetics. He is the coeditor of *Eskimo of Northwestern Alaska: A Biological Perspective* (with P. L. Jamison and F. A. Milan; Dowden, Hutchinson, and Ross, 1978).